DATE DUE			
AUG 26 1996			
JUN - 3 1997			
MAR 16 1998			
APR - 4 2000			
9-7			

Demco, Inc. 38-293

ENRICO FERMI
PHYSICIST

ENRICO FERMI PHYSICIST

EMILIO SEGRÈ

THE UNIVERSITY OF CHICAGO PRESS
Chicago and London

International Standard Book Number: 0-226-74472-8
Library of Congress Catalog Card Number: 71-107424
The University of Chicago Press, Chicago 60637
The University of Chicago Press, Ltd., London
© 1970 by The University of Chicago
All rights reserved
Published 1970
Printed in the United States of America

Contents

List of Illustrations / vii

Preface / ix

1 / Family Background and Youth / 1

2 / Apprenticeship / 25

3 / Professor at Rome / 46

4 / Emigration and the War Years / 101

5 / Professor at Chicago / 156

APPENDIXES

1 / Letters to Enrico Persico / 189

2 / Artificial Radioactivity Produced by Neutron Bombardment / 214

3 / Physics at Columbia University / 222

4 / The Development of the First Chain-Reacting Pile / 231

Notes / 240

Bibliography / 251

Index / 269

Illustrations

Facing page 1

Fermi solving a problem, around 1924

Following page 116

Fermi at the age of sixteen
Segrè, Fermi, and Persico at Ostia, 1927
The Rome Institute of Physics
The library of the Rome Institute
Fermi, Corbino, Trabacchi, Sommerfeld, and Zanchi, 1931
The Solvay Conference
Manuscript of the beta ray theory
D'Agostino, Segrè, Amaldi, Rasetti, and Fermi, 1934
The "Roman Sign": ionization chamber used for measuring radioactivity
The Fermi family landing in New York
Fermi and Oppenheimer
Los Alamos, summer 1945: Lawrence, Fermi, and Rabi
Fourth anniversary reunion, December 2, 1946
Target holder for cyclotron
Fermi lecturing

Preface

This is a biography of Enrico Fermi as a scientist. Physics was the essence of his life. Although he lived in a time of great human drama and, through his work, became a major actor in it, his personal involvement was with the intellectual adventure of scientific discovery. Other aspects of his life are excellently treated in Laura Fermi's book *Atoms in the Family*. Our points of view are different: in one case a devoted, loving wife; in the other, a friend, disciple, and fellow scientist.

I have occasionally tried to guess what I have been unable to ascertain; these guesses are identified where they appear. (Contrary to the impression of casual acquaintances, Fermi was not easy to approach.) My sources of information besides recollection—an untrustworthy one—are Fermi's documents and manuscripts deposited at the University of Chicago Library and the books I have cited in the text. Only a few of his letters are extant; Fermi did not write very much, and he preserved only a small part of his correspondence. If my correspondence with him is typical, his letters of the American period were almost all either on business matters or on scientific subjects that he later published. Some early correspondence of interest for the Rome period unfortunately was lost in a shipment from Italy on the ill-fated *Andrea Doria*.

I have freely used *The Collected Papers of Enrico Fermi* and the introductions therein. Mrs. Fermi, Edoardo Amaldi, and Enrico Persico have been kind enough to go over the manuscript of my book, for which I thank them wholeheartedly. Their help has been invaluable, but they cannot be held accountable for my writing. Mrs. R. Fowell patiently typed the manuscript, and my wife, E. H. Segrè, compiled the index. To them also I am very grateful.

Preface

Marie Curie once wrote *"En science, nous devons nous intéresser aux choses, non aux personnes* [In science we must interest ourselves in things, not in persons]," and the sentence could have been Fermi's; nevertheless, curiosity about the personality of great scientists seems to me legitimate and certainly is evidenced by many aspiring scientists. I have tried to satisfy this curiosity, within my limitations, in a spirit of objectivity.

ENRICO FERMI
PHYSICIST

Enrico Fermi Solving a Problem, Around 1924

1
FAMILY BACKGROUND AND YOUTH

Wednesday is a market day in Rome, and one Wednesday late in 1915 a fourteen-year-old boy, short and stocky, was making his way among the old book stalls. He had walked at a brisk pace the two miles from his home to hunt for physics books. The boy's expression was serious and determined and his manner grave. The First World War was raging, lengthy casualty lists were published, and the preoccupation with these dreadful events percolated to the younger generation. The boy had also recently experienced a grievous personal loss in the death of his brother, older by one year. Although he never talked about this loss, he was searching the bookstalls for something precious to him that would help him forget the wound. He found mainly tomes on theology (Rome is full of priests), discarded novels, and old art books; his search seemed a waste of time. Discouraged, he opened two old tomes, unpromising from the outside because the titles had disappeared from the covers, but, oh joy, they were full of mathematical symbols. They were on physics, exactly what he was looking for, and he read the title: *Elementorum physicae mathematicae* ..., by Andrea Caraffa of the Society of Jesus. The year of publication was 1840, and the volumes were in Latin; but they were better than nothing, and his tedious study of Latin would now find worthy use. Not in vain had he learned so much Latin grammar and always had the best grades in it. His weekly allowance, after some haggling,

Family Background

sufficed to buy this treasure. Thus did Enrico Fermi acquire his first physics book.

I heard Fermi speak for the first time about ten years later. He was telling astounding things—unthinkable to Father Caraffa in the nineteenth century—to an audience made up of some of the most famous mathematicians in Rome, to my own professors, and to some fellow students who were only a few years younger than the speaker. This was my first acquaintance with a future teacher and friend.

Twenty years later, at the dawn of a July morning in a New Mexico desert, Fermi and I witnessed the awesome sight of the first nuclear explosion, a direct consequence of his work.

In less than another ten years, in the chapel of the University of Chicago, a deeply saddened crowd paid its last tribute to a beloved friend, teacher, and colleague. The chaplain, having thought hard to find a text suitable to the man and the occasion, chose Saint Francis' *Canticle of the Creatures,* which reflected Fermi's deep love for nature:

> Altissimo onnipotente bon Signore, Tue so le laude,
> la gloria e l'onore et onne benedictione.
> Most high, all powerful, good Lord, yours are the
> praise, the glory, the honor, and all blessings.

Enrico Fermi was born in Rome on September 29, 1901, to a family that originally was composed of peasants from a rich agricultural region near Piacenza, in the Po Valley in northern Italy. Stefano Fermi, Enrico's grandfather, was the first member of the family who did not work the soil with his own hands. By hard, stubborn toil he had improved his social status and had obtained a modest position in the service of the Duke of Parma, one of the minor princes who ruled in the Po Valley after Napoleon's downfall and the Peace of Vienna. The Dukes of Parma were under strong Austrian influence and represented reactionary and autocratic ideas, but they seem to have been close to their subjects, who were few in number and concentrated in a small region. Stefano, we can surmise, was a hard, parsimonious man who combined ambition and intelligence with a purposeful narrowness of mind. Enrico in his infancy met him and long remembered his stern grandfather. Despite his hard work, Stefano remained relatively poor. At his death in 1905, at the age of eighty-seven, he owned a small house and some land in Caorso, near Piacenza. Stefano's wife, Giulia Bergonzi, born in 1830,

Family Background

was thirteen years younger than Stefano. She was a typical nineteenth-century Italian woman of the countryside, a type often described by Italian novelists. She was devoted to the church, had a large family, worked long hours in the house, and possessed all the domestic virtues. In the presence of others, at least, she always called her husband simply "Fermi." She was loved by her children and grandchildren, and spent her leisure time reading *I Promessi Sposi*, the classic catholic novel by Manzoni, and her book of devotions—but she never learned to write. The children were brought up religiously, and all except Alberto, Enrico's father, remained faithful to the church. As a grandmother, Giulia was sorry that Alberto's children had not followed the family's religious tradition, but she respected the will of the parents. When the grandchildren visited her, she always had little presents for them, of no trivial import to her extremely tight budget.

Alberto Fermi, the second son of Stefano, was born at Borgonure, near Piacenza, on April 3, 1857. He was brought up in an austere atmosphere, in which the desire for social and economic improvement conflicted with pressure toward early gainful employment. I do not know the extent of Alberto's formal education, but it is likely that he attended a technical high school. Such schools, instituted shortly after the unification of Italy, were intended to train young people for practical careers, and they were attended chiefly by boys who could not afford the expenses of higher education. They gave limited access to the university—namely, to schools of engineering—but in general, graduates of the technical high schools were not expected to continue their studies.

Alberto Fermi must have had some gainful occupation before 1882, but I have found no record of it. In 1882 he was hired by one of the private railroad companies of northern Italy, the Alta Italia, and started a career as a railroad employee, which he pursued successfully until his retirement in 1921. Because of mergers among Italian railroad companies, he was on the payroll of several of them and held administrative jobs that took him to various parts of Italy. In 1885 he was stationed in Naples, working for the Mediterranea Company. Later, probably in 1888, he was transferred to Rome. By 1890 he was an accountant, by 1898 an inspector, and in 1901 he received his first decoration, the title of Cavaliere, which at that time had not been debased by overuse. When the state acquired the railroad, Alberto became a government civil servant, and he ultimately

Family Background

reached a fairly high post, Capo Divisione, which in the civil administration corresponded to the rank of brigadier general in the army. Also he was awarded the decorations of commander in the honorary orders of Saints Maurizio and Lazzaro and the Crown of Italy. Alberto's rise to such posts, without an advanced education, indicates unusual abilities.[1] People who knew him described him as serious and taciturn, but he sang arias from Verdi's operas while shaving (thus giving Enrico his first "musical" education).

Fermi's mother, Ida de Gattis, was born at the other end of Italy, in Bari, on April 10, 1871, the daughter of an army officer. She was trained as a school teacher and taught in the elementary schools most of her life. She was considered unusually intelligent and able by her children and their friends. At the time of their marriage in 1898, Alberto was forty-one years old and Ida twenty-seven. Despite this disparity in age, however, the young woman became the stronger influence in the family. For ten years the couple lived in an apartment in Via Gaeta 19, near Rome's railroad station. In 1908 they moved to Via Principe Umberto 133, an apartment house in which Enrico spent his early youth.

The Fermis' apartments were located in quarters that had been hurriedly developed after the establishment of Rome as the capital of Italy in 1870. This event brought to Rome, sleepy in the last years of the temporal power of the popes, an extraordinary influx of workers, politicians, businessmen, and others, and the city's population grew from about 180,000 in 1870 to about 400,000 at the turn of the century. In accommodating all these people, Rome experienced a building boom that created entire new districts renowned chiefly for their ugliness—unimaginatively planned grids of rectangular blocks that showed little regard for the architectonic character of Rome. Inspired by a mechanical, uncomprehending imitation of the palaces of the late Renaissance, the false grandiosity of the buildings usually offered insufficient amenities and comforts, and the population that occupied these districts was composed mostly of immigrants from other parts of Italy, white-collar workers in the low echelons of the administration and small businessmen.

The social composition of the district in which the Fermis settled had important effects on the schools, especially on the secondary schools. Rome's *ginnasi-licei* were the humanistic schools that alone gave full access to the university; their programs of study were uniform, but attendance at one or another of these schools implied

Family Background

subtle but important differences in the education of the students. It also gave a particular stamp to their social position, similar to the school-tie distinctions in Great Britain. The school attended by Fermi enrolled mostly lower middle-class pupils.

The Fermis' first child was a girl, Maria, born in 1899. In 1900 a boy, Giulio, was born, and on September 29, 1901, a second boy, Enrico. The infants were sent to wet nurses in the country, a common practice at the time, and Enrico was returned to his family only when he was two and a half years old. He adjusted rapidly to his family and became very much attached to his parents and siblings. His mother had a stern sense of duty and discipline, but this was coupled with intelligence and great devotion to the family. She had a strong influence on her children and inspired them by example to a serious conception of life and work. Her love also had a deep unifying effect on the family. The children did not receive religious instruction, although they had been baptized in deference to the grandparents' feelings. Enrico Fermi's attitude to the church eventually became one of indifference, and he remained an agnostic all his adult life.

Enrico learned to read and write early, probably from his sister and brother. He soon displayed a prodigious memory, which he exercised by learning excerpts from Ariosto's *Orlando Furioso,* a heroic-comic poem written around 1530. Its subjects are chivalry and the love madness of Roland, and the story is set in the time of the Moors' invasion of Spain. It can be quite amusing and was greatly appreciated by Galileo. I can only wonder why it so fascinated Enrico, but his sister remembered that he recited it with great glee. As an adult, he did not seem to remember Ariosto very well, certainly not to the extent to which he remembered Dante.

Fermi began attending the then strictly secular public school at the customary age of six. He did well, getting high marks, and soon showed unusual mathematical ability.

Fermi never told me how he first became acquainted with mathematics; it is possible that a friend of his father introduced him to the subject. Because the Italian railroads had rather high standards of technical proficiency, some of the colleagues of Alberto Fermi were interested and well read in mathematics. At any rate, Fermi told me that one of his great intellectual efforts was his attempt to understand—at the age of ten—what was meant by the statement that the equation $x^2 + y^2 = r^2$ represents a circle. Someone must have stated the fact to him, but he had to discover its meaning

Family Background

by himself. Thus at ten he had made real efforts to understand abstract problems.

Ten also was the age at which a boy would leave the elementary school and enter the ginnasio for five years, to be followed by three years at the liceo, in preparation for the university. The ginnasio and liceo emphasized Italian, Latin, and Greek, and offered instruction in history, geography, mathematics, physics, natural history, philosophy, and French—all of which was compulsory. The work load was heavy for average students, and the instruction was not superficial.

The exaggerated emphasis on literary subjects was a weak point of this educational system, which often tended to stress ability in rhetoric and a rather pompous literary style. Fermi's professor in Italian at the liceo, Giovanni Federzoni (the father of a future leader of the Nationalist and Fascist parties in Italy), knew Dante well, and the reading of the *Divine Comedy* seems to have made a strong impression on Fermi, who knew many long stretches of the poem by heart and in later life occasionally quoted them with light irony, as if he were making fun of himself. On the other hand, Fermi did not acquire any particular stylistic refinement of his own, and for many years his writing was strictly utilitarian. He expressed himself with clarity, because his thinking was clear, but in his early papers and in his *Introduzione alla fisica atomica* [Introduction to atomic physics], written in 1927, there was no effort to polish his literary style or even to avoid a rather flat and careless use of language. In later years, however, he became very particular, almost pedantic, about precision of expression, and the difference in style between his early papers and those of his maturity is easily discernible.

I do not remember that Fermi ever referred to his Latin studies, but many years later, at Los Alamos, I complained one day that the night before I had dreamed of a final liceo examination in Greek, to which he remarked that he had suffered similar nightmares.

His formal scholastic record continued to be excellent; he was easily the best student in his class. Because of his habits of order and discipline, he had considerable free time, most of which he seems to have devoted to scientific studies.

His closest friends were his sister and brother, and with the latter he built electric motors and other mechanical or electrical toys. Giulio, one year older than Enrico, was by all indications just as bright as Enrico, but he had a more artistic and warmer tempera-

Family Background

ment and was his mother's favorite. The two brothers, who always played together, formed a little world that was somewhat closed to outsiders but perfectly sufficient for them.

On January 12, 1915, when Enrico was not yet fourteen, a very severe blow fell suddenly and unexpectedly on the Fermi family. Giulio developed an abscess in his throat and had to be operated on. Nobody thought there was any danger; then, unexpectedly, he died under anesthesia. This tragedy completely devastated Mrs. Fermi, who became deeply melancholic, often cried for hours, and could not help the family face the sorrowful situation. Enrico also was extremely affected, but his rather taciturn and introverted character prevented him from showing his feelings. He had lost his only brother and his best and almost only playmate. Study was one of Enrico's consolations at this sad time, and he gave himself to even deeper and more difficult study.

Soon he was fortunate in finding a new friend, Enrico Persico, who had been a schoolmate of Giulio, but not a particular friend of either brother, because Giulio and Enrico had kept so much to themselves. Enrico Persico, of whom we shall hear more later, was born in Rome in 1900. He too was interested in science, and he and Fermi were to become the first two professors of theoretical physics in Italy. Persico wrote as follows of the early times of his friendship with Fermi.

When I first met him he was fourteen years old—I noted with surprise that I had a schoolmate not only "strong in science," as one used to say, but also with a type of intelligence entirely different from that of other boys whom I knew and whom I considered good students and intelligent.

We formed the habit of taking long walks together, crossing the city of Rome from one side to the other, discussing all kinds of subjects with the brashness characteristic of youth. But in these adolescent talks Enrico brought a precision of ideas, a self-assurance, and an originality which continually surprised me. Furthermore, in mathematics and in physics he showed knowledge of many subjects well beyond what was taught at school. He knew these topics not in a scholastic fashion, but in such a way that he could use them with extreme facility and familiarity. For him, even at this time, to know a theorem or a law meant chiefly to know how to use it.

In reflecting on the feeling of surprise and admiration which Enrico's intelligence produced in me, a boy of almost the same age, I asked myself whether I ever applied to him the word "genius." Probably not, because for children, and perhaps for most grown-ups also, this word as a rule is associated less with a quality of mind than with the social image of

Family Background

a person old, famous, and inaccessible. The quality of the mind of my friend was something too new for me to be able to name.[2]

Some of the long walks in Rome were visits to the market of Campo dei Fiori, the very market in which Fermi was observed at the beginning of this book. Campo dei Fiori is a square in the heart of the Renaissance center of Rome whose well-preserved classical palaces, such as the Palazzo Farnese by Michelangelo and the Cancelleria by Sangallo, still bear testimony to the Rome of 1500. In this square Giordano Bruno was burned at the stake, a reminder of the darker aspects of the same period. Fermi, however, was not looking for historical inspiration but simply for secondhand books. It was here that he found the textbook by Caraffa. The book is a fairly good text at the university level on mechanics, optics, acoustics, and astronomy as these were known in 1840. It also includes a mathematical introduction in which calculus is explained in the style used in the eighteenth century—lacking rigor but possessing a certain algebraic virtuosity. Fermi must have studied the treatise very thoroughly, because it contains marginal notes, corrections of errors, and several scraps of paper with notes in Fermi's handwriting. The two volumes gave Fermi a solid foundation for his future studies.

About this time Fermi met another person who was to be important for his development, a colleague of his father, Adolfo Amidei. Amidei was an engineer (*ingegnere*)—that is, a man with advanced technical training obtained by university study—and Fermi several times mentioned to me Amidei. In 1958 I asked the old gentleman to give me an account of his recollections of the young Enrico Fermi. His account, contained in a letter which is in my possession, is as follows:

In 1914 I was principal inspector [a position lower than that of chief inspector] of the Ministry of Railways, and my colleague was chief inspector Alberto Fermi. When we left the office we walked together part of the way home, almost always accompanied by the lad Enrico Fermi (my colleague's son), who was in the habit of meeting his father in front of the office. The lad, having learned that I was an avid student of mathematics and physics, took the opportunity of questioning me. He was thirteen and I was thirty-seven.

I remember very clearly that the first question he asked me was: "Is it true that there is a branch of geometry in which important geometric properties are found without making use of the notion of measure?" I replied that this was very true, and that such geometry was known as

Family Background

projective geometry. Then Enrico added: "But how can such properties be used in practice—for example, by surveyors or engineers?" I found this question thoroughly justified, and after having tried to explain some properties that had very useful applications, I told him that the next day I would bring him—as I did—a book on projective geometry by Professor Theodor Reye[3] that included an introduction which, in a masterful, artistic style, succeeds in explaining the usefulness of the results of this science.

After a few days Enrico told me that besides the introduction he had already read the first three lectures and that as soon as he finished the book he would return it to me. After about two months he brought it back and, to my inquiry whether he had encountered any difficulties, replied "none," adding that he had also demonstrated all the theorems and quickly solved all the problems at the end of the book (there are more than two hundred).

I was very surprised, and since I remembered that I had found certain problems quite difficult and, since they would have taken me too much time, had given up trying to solve them, I wanted to verify that Enrico had also solved those. He gave me the proofs.

Thus it was certainly true that the boy, during the little free time that was left to him after he had fulfilled the requirements of his high school studies, had learned projective geometry perfectly and had quickly solved many advanced problems without encountering any difficulties.

I became convinced that Enrico was truly a prodigy, at least with respect to geometry. I expressed this opinion to Enrico's father, and his reply was: Yes, at school his son was a good student, but none of his professors had realized that the boy was a prodigy.

I then learned that Enrico studied mathematics and physics in secondhand books that he bought in the Campo dei Fiori, hoping to find a treatise that would scientifically explain the motion of tops and gyroscopes, but he could not find an explanation, and so, mulling the problem over and over again in his mind, he succeeded in reaching an explanation of the various characteristics of the mysterious movements by himself. Then I suggested to him that to obtain a rigorous explanation it was necessary to master a science known as theoretical mechanics; but in order to learn it he would have to study trigonometry, algebra, analytical geometry, and calculus, and I advised him not to try the problems of tops and gyroscopes since he would be able to solve them easily once he had mastered the fields that I had outlined. Enrico was convinced of the soundness of my advice, and I supplied him with the books I thought most suitable for giving him clear ideas and a solid mathematical base.

The books which I lent him and the dates of the loans are as follows:

In 1914, for trigonometry, the treatise on plane and spherical trigonometry by Serret.

In 1915, for algebra, the course on algebraic analysis by Ernesto Cesàro, and, for analytical geometry, notes from lectures by L. Bianchi at the University of Pisa.

Family Background

In 1916, for calculus, the lectures by Ulisse Dini at the University of Pisa.

In 1917, for theoretical mechanics, the *Traité de mécanique* by S. D. Poisson.

I also deemed it appropriate for him to study the *Ausdehnungslehre* by H. Grassmann, which has an introduction on the operations of deductive logic by Giuseppe Peano. These books were lent to him in 1918. I thought it appropriate because it was my opinion that the *Ausdehnungslehre* [similar to vector analysis] was the most suitable tool for the study of different branches of geometry and theoretical mechanics, because in this system of calculus the operations are not performed on numbers [coordinates] which define the geometrical objects but on the objects themselves, thus obtaining formulae which are simple and of easy interpretation. Furthermore, this calculus contains, as a particular case, analytical geometry, and is especially suited to the study of theoretical mechanics and graphical statics.

Enrico found vector analysis very interesting, useful, and not difficult. From September 1917 to July 1918 he also studied certain aspects of engineering in books that I lent him.

In July, 1918, Enrico received his diploma from the liceo (skipping the third year), and thus the question arose whether he should enroll at the University of Rome. Enrico and I had some long discussions on this subject.

First of all I asked him whether he preferred to dedicate himself to mathematics or to physics. I remember his reply, and I transcribe it here literally: "I studied mathematics with passion because I considered it necessary for the study of physics, *to which I want to dedicate myself exclusively.*" Then I asked him if his knowledge of physics was as vast and profound as his knowledge of mathematics, and he replied: "It is much wider and, I think, equally profound, because I've read all the best-known books of physics."[4]

I had already ascertained that when he read a book, even once, he knew it perfectly and didn't forget it. For instance, I remember that when he returned the calculus book by Dini I told him that he could keep it for another year or so in case he needed to refer to it again. I received this surprising reply: "Thank you, but that won't be necessary because I'm certain to remember it. As a matter of fact, after a few years I'll see the concepts in it even more clearly than now, and if I need a formula I'll know how to derive it easily enough."

In fact, Enrico, with his marvelous aptitude for the sciences, possessed an exceptional memory.

I then considered that the proper moment had arrived to present a project that I had already considered for him for a year, that is, from the time when I advised him—and he immediately followed my suggestion—to study the German language, because I foresaw that it would be very useful for reading scientific publications in German without having to wait until they were translated into French or Italian.

Family Background

My plan was this: Enrico ought to enroll not at the University of Rome but at the University of Pisa. First, however, he would have to win a competition to be admitted to the Scuola Normale Superiore at Pisa and also attend the courses in that school. This would not have caused additional expenses because the Scuola Normale gave room and board to its fellows. Moreover, it would give Enrico several advantages:
1. An excellent library was available to its students.
2. The subjects taught there supplement those taught at the university, or sometimes are entirely new.
3. Having learned German and knowing French, Enrico could correspond with professors of the École Normale in Paris or with German scientists who every year published the new progress in physics, which was moving rapidly (quantum theory, relativity, etc.).
4. He could study with greater comfort and peace, inasmuch as a very depressing atmosphere prevailed in his home after Giulio's death.

Enrico recognized at once the soundness of my plan and decided to follow it, even though he knew that his parents would be opposed to it. Then I immediately went to Pisa to obtain the necessary information and the program for the competition for the Scuola Normale Superiore, and returned to Rome to examine it minutely with Enrico. I ascertained that he knew mathematics and physics perfectly, and I expressed my conviction that he would not only win the competition but would be the first among the applicants, as in fact he was.

Enrico's parents did not approve of my plan, for understandable human reasons. They said: "We lost Giulio and now we are to allow Enrico to leave us to study at Pisa for four years while there is an excellent university here in Rome. Is this right?"

I used the necessary tact to persuade them, step by step, that their sacrifice would open a brilliant career for their son, and they finally agreed that my program should be carried out. Thus, as Enrico's wife wrote in her book *Atoms in the Family,* "at the end, the two allies—Fermi and Amidei—carried the day."

Enrico's statements, which I quoted above, are accurate because I wrote them down the same day they were exchanged.

Amidei's letter is an important document and, although written many years after the events it describes, deserves to be reported in its entirety for the light it sheds on both its author and its subject.

At the time that Fermi was studying physics in books he was also acquiring various skills by performing experiments. He had started by building electric motors, airplane models, and electrical and mechanical toys with his brother, as noted before; later, he and Persico increased their knowledge of physics and refined the subjects and the art of their experiments. I know, for instance, that they tried to measure accurately the acceleration of gravity at Rome, the density of tap water, and similar quantities. They wrote down the results of some of their work, and their notes include impressive discussions of the errors and of the precision of the measurements. At about this time they also made a memorable study of the top, or gyroscope, whose behavior gave a great deal of trouble to the young scientists. After great efforts, Fermi succeeded in developing a theory that was satisfactory to himself and his friend, but the details of this work are lost.

Fortunately for the biographer, Persico preserved the letters he received from Fermi between 1917 and 1926. He has very kindly given me photographic copies of these documents, which present a vivid picture of the life and personality of the young Fermi. The correspondence is reproduced in its entirety in Appendix 1.

In the first communication—a postcard dated September 7, 1917—Fermi said he spent every morning studying in the public library of Rome. In preparation for the competitive examination for admission to the Scuola Normale Superiore at Pisa, he was studying systematically the large physics treatise by Chwolson, then a standard advanced text. This was approximately the time at which he also mastered the *Mécanique* of S. D. Poisson (mentioned in Amidei's letter). This book impressed him greatly; he recommended it repeatedly to Persico and even, many years later, to me. Obviously, he assimilated its contents very thoroughly. Almost forty years later, when he gave me a subtle proof of the composition of forces, I was rather surprised because it did not seem at all in the familiar Fermi style. About that time I found a secondhand copy of an old edition of Poisson and gave it to Fermi as a present. Browsing in it, I had found that Fermi, completely forgetful of its origin, had reproduced Poisson's proof of the theorem.

On November 14, 1918, when Fermi presented himself at the competition (held in Rome) for admission to the Scuola Normale Superiore, he was given the theme for his essay: "Characteristics of Sound." After a few introductory remarks, the essay (preserved in the archives of the Scuola Normale) set forth the partial differential equation of a vibrating rod, which Fermi solved by Fourier analysis, finding the eigenvalues and the eigenfrequencies. The entire essay continues on this level, which would have been creditable for a doctoral examination and was totally unexpected from the usual competitors—bright boys out of the liceo, with a thorough preparation at the high school level.[5]

Family Background

The examiner was Giuseppe Pittarelli, professor of descriptive geometry at the University of Rome, a competent mathematician, a good amateur painter, and a kind gentleman. He was so surprised by the level and the excellence of the essay that he decided he had to meet its author. He called Fermi in to speak to him, although this was not prescribed by the rules. The interview ended with Pittarelli's telling Fermi that in his long career as a professor he had never seen anything like this, that Fermi was a most extraordinary person and was destined to become an important scientist. As for the competition, he added, Fermi could be sure of winning because it was unthinkable that another candidate could match his accomplishment. Fermi himself told me this story many years later, with obvious pleasure, and also with gratitude to Pittarelli for his encouragement and praise, which gave Fermi self-confidence. In the fall 1918 Fermi left Rome for the Scuola Normale in Pisa, where, except for vacations, he was to spend the next four years. From a modern metropolis—even in 1918 Rome had such a character—Fermi moved to a quiet, slumbering town whose days of splendor had been the Middle Ages.

Pisa, which originally stood on the seashore, now lies about ten miles from the sea on the Arno River, in flat country surrounded by rich agricultural land. After a period of glory as a naval republic, Pisa around the year 1000 became one of the great cities of the late Middle Ages and an important center for the development of the new art leading to the Renaissance. The Duomo (cathedral) was started around 1063, and its campanile (the Leaning Tower) was built one hundred years later. At the end of the thirteenth century —the time of Dante—Pisa's political and military power diminished as a consequence of internal dissensions and naval defeats at the hands of Genoa, but its artistic and cultural development continued. The University was founded in 1343, and when the Medici of Florence conquered Pisa in the next century they fostered and enhanced the university. From the sixteenth century on it was the main university in Tuscany, and it attained considerable splendor. Within a relatively short time, Andrea Cesalpino, Gabriello Fallopio, Galileo Galilei, and Marcello Malpighi taught there. In the nineteenth century the university's political consciousness increased, and during the Risorgimento the institution became a center of liberal political activity. Student life at Pisa during this period has been de-

Family Background

picted repeatedly, notably by the poet Giuseppe Giusti, who studied there around 1830. Some of its rather romantic and bohemian atmosphere is still present, albeit tempered by the conservatism of the Italian middle class and the caustic and skeptical spirit of Tuscany.

After the unification of Italy the University of Pisa experienced a new period of splendor, attracting many of the best literary and scientific talents of Italy. Distinguished mathematicians, such as Enrico Betti, Ulisse Dini, and Luigi Bianchi, and the physicist Ottaviano Mossotti are familiar names to students of their disciplines. Similarly, Pasquale Villari, Alessandro d'Ancona, and Giovanni Gentile are renowned in the humanities. The great Bernhard Riemann, trying to regain his health in a better climate than that of Göttingen, spent the years 1858 to 1860 in Pisa and participated in the mathematical activity of the university.

The original purpose of the Scuola Normale Superiore, which was attached to the university and had been founded in 1810 by Napoleon as a branch of the École Normale Supérieure of Paris, was the preparation of high school teachers and the promotion of advanced study and research. The Scuola Normale, closed after Napoleon's downfall but reopened by the Grand Duke of Tuscany in 1846 as a sort of elite college attached to the university, more or less shared the fortunes of the latter. It has had a number of brilliant alumni, such as the poet Giosuè Carducci, the mathematicians Gregorio Ricci-Curbastro, Vito Volterra, Guido Fubini, Guido Castelnuovo, Francesco Severi, and the philologists Michele Barbi, Pio Rajna, and Girolamo Vitelli. Such persons, most of them of permanent importance in their fields of study, constituted a sizeable portion of Italy's intellectual leaders and indirectly conferred great prestige on their alma mater. Thus the Scuola Normale, as well as the similar Collegio Ghislieri in Pavia, with a restricted student body that was selected nationally in severe and impartial competitions, has been a major source of talent for Italy.

The Scuola Normale is housed in the Palazzo dei Cavalieri, on a stately square not far from the famous cathedral and its Leaning Tower. Galileo, when he was a student at Pisa, had almost the same view of that part of the city as we have today. In 1918 the town of fifty thousand, despite its ten thousand university students, was rather sleepy, and with its historical and artistic wealth, it had somewhat the character of a living museum. The palace of the Scuola Normale, grandiose on the outside, was cold in winter, as all Renaissance

Family Background

palaces are. In Fermi's time the heating was provided by charcoal that burned in *scaldini,* portable ceramic braziers, and there was no hot running water. The students' rooms, reminiscent of monastic cells, were large but sparsely furnished. The food was good and abundant, but the menu varied little; service was provided by male stewards who served the meals and took care of the rooms.

There were about forty students in all, divided into two sections: one working in the humanities, the other in mathematics and science. A professor of Greek was the resident director of the school. The students, all of high ability and seriously motivated, concentrated on their studies and formed a rather quiet group. To Fermi and his friends, sports consisted primarily of hiking in the Apuan Alps, not far from Pisa. These mountains, famous for the quarries that yield the celebrated Carrara marble, are not very high, but because they start at sea level and are exceedingly steep and rugged, they offer plenty of strenuous exercise for climbers. The plains at the base of the mountains are lush with rich farms and are well suited for bicycling.

Fermi at first felt strange in the new surroundings. Probably he was homesick, but after a few weeks he overcame his despondency and regained his usual composure. Not long after his arrival at Pisa he struck up a friendship with Franco Rasetti, a physics student at the university (though not a fellow of the Scuola Normale) who lived with his family in Pisa.

The Rasettis had substantial agricultural holdings, and Franco's father had taught in the agricultural extension; his mother was the sister of a well-known physiologist, Galeotti, and a good amateur painter. Rasetti, an only child, lost his father while he was at the university, and his mother concentrated all her affection on Franco. The boy developed an extraordinary intellectual bent but had little interest in people. With a collector's instinct, he was particularly attracted to natural science; he gathered and classified insects, plants, and the like—with the minutest care and with deep scientific insight. His memory, like Fermi's, was prodigious, and he was interested and successful in many branches of science. Over the years, in addition to physics, he cultivated entomology, embryology, botany, and paleontology with a proficiency equal to that of professionals. According to Laura Fermi, Rasetti registered in physics at the university rather than biology, which might have been nearer his deep-seated interests,

Family Background

"because it was not easy for him to understand physics and he wanted to prove to himself that he could overcome any difficulty."[6] Rasetti, at any rate, became a very close friend of Fermi, who often visited him at his home and frequently ate there on Sunday, relieving the monotony of the school's fare.

Fermi also became active in an "anti-neighbor society," organized by Rasetti, whose aim was to pester people with pranks. The tricks the "society" played are typified by the following: At that time in Italy there were on suitable street corners public urinals that were built in such a way that a user stood close to a pool of water. Members of the society would sneak up behind a man and throw small pieces of metallic sodium into the water, and then watch from a safe distance as the sodium caught fire and exploded. In other capers they would surreptitiously lock with small padlocks the coats or overcoats of people diverted in conversation by accomplices, or would fight mock duels on the rooftops of the city.

Study, however, was the real occupation. Fermi carefully read Poincaré's *Théorie des tourbillons,* Appell's *Mechanics,* and Planck's *Thermodynamics.* But his study was not limited to reading; he solved innumerable problems, self-given or found in books. Sometimes he would propose a problem by letter to his friend Persico or give him advice with the air of a big brother—despite Persico's being one year older. He must have spent a tremendous amount of concentrated thought on the classics of physics, thoroughly assimilating them and distinguishing the important concepts and methods from the purely elegant ones, with unerring judgment. Many years later he could instantly produce proofs for all the important theorems of physics, proofs which often were almost identical to those in the books he had studied at Pisa, sometimes even in the wording. As in the case of Poisson's book, he had completely forgotten the source of his knowledge. Apparently in his later years Fermi mentally rehearsed chapters of physics as a director rehearses a symphony. He would do this on long cross-country drives or similar occasions.

During the summer vacation of 1919 he joyfully returned to Rome, his family, and his friend Persico. He also went to Caorso, to the house of his grandfather. Then he decided to organize his knowledge of physics, and he used the remainder of his summer vacation to do this. In Rome and Caorso, he filled with his notes a leather-bound booklet which is preserved among Fermi's papers at the Uni-

Family Background

versity of Chicago. This means that he kept it all his life and did not leave it in Italy at the time of his emigration (although I must also say that I had never seen the booklet before Fermi's death). The booklet, divided into parts and written in pencil almost without erasures, gives us a clear picture of Fermi's scientific preparation and intellectual progress in 1919, when he was seventeen and, later, eighteen years old. The first twenty-eight pages contain a summary of analytical dynamics and are dated Caorso, July 12, 1919. In it he develops the theory of Hamilton and Jacobi, treating very advanced topics with extreme conciseness and clarity. There are no indications of his sources of information, but very likely they are the works of Poincaré and Appell. The next pages are on the electronic theory of matter; they are dated Rome, July 29, 1919, and they contain a résumé—as usual very concise—of the subject. In these pages he treats Lorentz's theory of electrons, relativity, the blackbody theory, diamagnetism, and paramagnetism. For this part there is a bibliography that lists several of his sources on this subject, including Richardson's *Electron Theory of Matter,* which we know he studied carefully.

Bohr's first papers on the hydrogen atom also are mentioned, although at that time they were little known or appreciated in Italy. The following section, of nineteen pages, is dated Rome, August 10, 1919, and contains in greater detail the blackbody theory according to Planck. This is followed by an extensive bibliography on radioactive substances and their decay, taken from Rutherford's *Radioactive Substances and Their Radiations.* There are no comments, and it is dated Caorso, September, 1919. The following chapter, from pages 81 to 90, is devoted to Boltzmann's H-theorem and kinetic theory, and is dated Caorso, September 14, 1919. We find the usual succinct but clear exposition of theory, with some applications. The method used for establishing the H-theorem is the same as that used by Boltzmann, which involves a detailed analysis of all collisions.

The booklet, totaling 102 pages, concludes with two bibliographies taken from Townsend's book on gas discharges. They deal with electrical properties of gases and photoelectricity. The last notes are dated Rome, September 29, 1919 (his eighteenth birthday) and are followed by a table of contents.

The booklet shows many of Fermi's characteristics in an embryonic stage. The choice of material is made with surprising discrimination, especially if one considers the author's age and the fact that he was virtually self-taught. Another characteristic is that Fermi, although never repulsed or frightened by mathematical difficulty, does not seek elegant mathematics for its own sake. Whether a theory is mathematically easy or difficult does not seem to concern him; the important point is whether it illuminates the essential physical content of the situation. If the theory is easy, so much the better, but if difficult mathematics is necessary, he is quickly resigned to such use. One also notices differences between the sections in which the logical structure of the subject predominates over its experimental content and those of a more empirical character. In the first, one perceives the master's hand; in the second, one sees the lack of experience and critical evaluation of the many papers quoted. All told, it is surprising that after one year of university work a student could put together such a booklet, which would be very creditable even for a teacher with a long educational career.

Fermi's unusual abilities were promptly recognized by most of his professors and fellow students. At the Scuola Normale, the other fellows—both in science and humanities—knew that he was a truly exceptional man. The director of the physics laboratory was Professor Luigi Puccianti, born in Pisa in 1875, a gifted man with a keen mind, but lazy. When he was a young man he had done interesting work on the anomalous dispersion of alkali vapors, but by the time Fermi came to Pisa, Puccianti had ceased to do original experimental research, although he remained interested in a critical understanding of classical physics, especially magnetism. Puccianti very soon saw that he had little to teach but much to learn from his student Fermi. He acknowledged this with the utmost candor and many times asked his student to "teach me something."

Fermi, fully aware of his preeminence, and without false modesty, told Persico in January 1920: "At the physics department I am slowly becoming the most influential authority. In fact, one of these days I shall hold, in front of several magnates, a lecture on the quantum theory, for which I am always a great propagandist." He was then a third-year student, eighteen years of age. He had mastered Sommerfeld's *Atombau und Spektrallinien*, and probably better than anybody else in Italy felt at home with the old Bohr-Sommerfeld quantum theory. This excellence in theoretical physics, however, did not distract him from serious and conscientious study of other required subjects, even those of little interest to him. He thought

Family Background

nothing of studying chemistry for six hours a day when this was called for. However, Rasetti described to me many times how he and Fermi often performed chemical analysis in a peculiar way: the laboratory was poorly equipped; the two friends were not inclined to do the slow, patient work that was required, and they found a shortcut. The substances to be analyzed were given to them as powders, and instead of dissolving them and proceeding according to the rules, they would put a sample of the unknown substance under a microscope in the physics department. A mixture, axiomatically, could contain only common, cheap constituents, because these were all the chemistry department could afford for instruction. Often, a good look at the powder would indicate its constitution, and then they would write an elaborate "step by step" report of a systematic analysis they had never performed.

I have no specific information on Fermi's studies in the summer of 1920, but his first letter to Persico after the beginning of the third school year in Pisa, dated November 29, 1920, contains a surprise: it is written in German. Obviously, Fermi had been practising this language, which at that time was of paramount importance for all students of physics (Sommerfeld's *Atombau und Spektrallinien* was the bible of atomic physics, and *Zeitschrift für Physik* probably was the most important journal). Fermi's German was good for a person who had never been in a German-speaking country, a few grammatical errors notwithstanding, and he wanted to show off to his friend. The ponderous style was that of the scientific literature.

Fermi could read French easily, a language required in high school and easy for an Italian to learn. He also could read scientific English, without difficulty, as is shown by the books he listed in his notes. In subsequent years Fermi came to know German well and spoke it fluently, and when he emigrated to the United States he made a very deliberate effort to improve his English.

With the end of the scholastc year 1919–20, Fermi had spent two years at Pisa, and having finished all of the preparatory courses, his status as a student changed. He was accepted into the physics laboratory in a position similar to that of an American graduate student. Thereafter he had only to take advanced specialized courses and prepare a dissertation for the doctor's degree, usually conferred at the end of the fourth year. Rasetti, one of the "collaborators" mentioned by Fermi in a letter to Persico dated November 29, 1920, described the situation as follows:

Family Background

In the fall of 1920, three students, Enrico Fermi, Nello Carrara, and Franco Rasetti, were admitted to the physics department at Pisa. Owing to the recently terminated First World War, no older students enjoyed a similar status; hence the three, aged nineteen, represented the entire "graduate" group at that time. Professor Luigi Puccianti, director of the physics laboratory, allowed them freedom of initiative to a degree seldom granted to students in Italy or elsewhere. They were allowed to use the research laboratories at all times, received keys to the library and instrument cabinets, and were given permission to try any experiment they wished with the apparatus contained therein. Carrara and Rasetti, who in the previous year had come to recognize Fermi's immense superiority in the knowledge of mathematics and physics, henceforth regarded him as their natural leader, looking to him rather than to the professors for instruction and guidance.

A few weeks were happily spent opening the instrument cabinets, guessing at the possible use of the apparatus, and sometimes trying simple experiments. It turned out that the laboratory was best equipped for research in spectroscopy, a field to which Professor Puccianti had brought important contributions. Some X-ray equipment also was available, although of the type designed for lecture demonstration rather than research.

Fermi, after much reading of the pertinent literature, decided that X-rays were the field that offered the best chance for original research, and suggested that all three learn some of the technique. The tubes available were of the gas-filled type and were operated by a large induction coil, using a spark gap as rectifier. The first task that Fermi set for the group was to produce a Laue photograph, and after several attempts and mishaps, one was obtained. Although the poor collimation of the beam and the imperfect orientation of the crystal had contributed to produce an unprepossessing Laue pattern, the three students were thrilled by the result.

It soon appeared that sealed tubes were not fit for research, and the experimenters decided to build their own tubes. The glass part was made to specification by a glass blower, while the physicists had to seal windows and electrodes. No diffusion pumps were available; hence the tubes were evacuated by means of rotary mercury pumps of the Gaede type. Considerable time was spent before these tubes could be satisfactorily operated, but eventually the K-radiations of several elements were obtained and observed by Bragg reflection.

Early in 1922 Fermi decided to undertake some original work with the X-ray technique he had been learning, in order to prepare a dissertation. Since at that time he had already published or at least completed several important theoretical papers, it may be asked why he did not present a theoretical thesis. It must be explained that at that time in Italy theoretical physics was not recognized as a discipline to be taught in universities, and a dissertation in that field would have been shocking —at least to the older members of the faculty. Physicists were essentially experimentalists, and only an experimental dissertation would have

Family Background

passed as physics. The subject nearest to theoretical physics, mechanics, was taught by mathematicians as a field of applied mathematics, with complete disregard for its physical implications. These circumstances explain why such topics as the quantum theory had gained no foothold in Italy; they represented a "no man's land" between physics and mathematics. Fermi was the first in the country to fill the gap.

However, there is not the slightest doubt in the writer's mind that Fermi's experimental activities in his third and fourth university years were not prompted by the convenience of presenting an acceptable dissertation. He obviously enjoyed experimental work as much as theoretical abstraction, and especially the alternation of the two types of activities. He was from the first a complete physicist for whom theory and experiment possessed equal weight, even though for many years his fame was chiefly based on the theoretical contributions. But he never for a moment was one of those theoreticians who, to use a joking expression later much in use among the Rome group, "could not tell steel from aluminum."[7]

Because of long neglect, the laboratory in which the young men worked was in very poor condition. It lacked what even at that time were normal facilities in good European laboratories, and much of its apparatus had been "cannibalized"—their parts disassembled for use in other instruments. Probably it was at Pisa that Fermi developed the habit of constructing everything with his own hands, a habit he later carried to strange excess. (I remember, for instance, that in Rome, despite the objections of his friends and associates, he spent much time and money constructing capacitors that should have been bought commercially.) The other personnel of the physics laboratory were Dr. Giovanni Polvani (Puccianti's first assistant), who became a good friend of Fermi, and two other assistants, Drs. Mariano Pierucci and Miss Anna Ciccone.

In January 1921 Fermi published his first paper, "On the Dynamics of a Rigid System of Electrical Charges in Translational Motion" (*FP* 1). This subject is of continuing interest; Fermi pursued it for a few years, and even now it occasionally appears in the literature.

In November 1921 Persico obtained his doctor's degree, and Fermi congratulated him in a special letter that has a crude drawing of two hands clasped in a congratulatory handshake. "The illustration on the side," he wrote, "represents what I would do if I were in Rome. I don't know whether you will be able to interpret it, because true works of art are always difficult to understand."

Most of 1922 was devoted to Fermi's preparation of his disserta-

Family Background

tion on X-ray diffraction by bent crystals and the images obtainable by this method. At the same time, however, he also had to prepare another separate dissertation for the Scuola Normale. The subject for the latter was a theorem on probability and its astronomical application. The university dissertation was discussed on July 7, 1922, and the doctor's degree was conferred *magna cum laude*. Soon thereafter he obtained his diploma from the Scuola Normale, and we find the unhappy history of the dissertation for the Scuola Normale in the letters of May 25 and June 2, 1922, addressed to Persico. Professor G. Polvani, a witness to the event, says that mathematicians leveled some criticisms at Fermi's calculations. The dissertation, "A theorem on probability and some of its applications," remained unpublished —certainly by Fermi's choice—until it was found in the archives of the Scuola Normale and was published as in the *Collected Papers* (*FP* 38*b*).

Meanwhile, something happened that was more important than either dissertation. In 1922, while still in Pisa, Fermi had obtained a remarkable result in general relativity by showing that, in the vicinity of a world line, space is Euclidean. This theorem is Fermi's first accomplishment of permanent value; it is contained in the paper "On the Phenomena Occuring near a World Line" (*FP* 3).

Different physicists love different aspects of their science. Some strive only for fundamental general principles, some hunt for new phenomena, some love precision measurements, some develop instruments or techniques. All these endeavors are not mutually exclusive; in fact they often supplement each other, and all are necessary to the progress of physics. The choice the physicist habitually makes is often described by the word "taste" or "style." By 1922, when Fermi left Pisa, he had developed his own style.

Fermi, as has been shown, was almost entirely self-taught; all that he knew he had learned from books or rediscovered by himself. He had found no mature scientists who could guide him, as he would have found at that time in Germany, Holland, or England, and he did not personally know any older scientist with whom he could compare himself. He knew that he was better than those around him, but this, he also knew, meant little, because these men were not in the forefront of active science. Moreover, although he had come in contact with eminent mathematicians, and although he was proficient in pure mathematics, he was not a professional mathematician and was little attracted to that discipline.

Family Background

In his choice of problems, therefore, he had of necessity been completely independent, except for his reading of the literature. It was characteristic of the young Fermi to prefer the analysis of definite phenomena whose explanation required a recondite use of known principles, imagination, and subtlety. He did not strive for major syntheses—or perhaps he was unable to accomplish them; often he was content with phenomenological theories. He never disdained the study of a given particular problem, but, remarkably, it always turned out that his work became a paradigm for further developments. Even in his major theoretical triumphs, the statistics and the theory of beta decay, he stays very close to particular problems. I believe that Fermi's inclination toward concrete questions verifiable by direct experiment was due, at least in part, to his desire to check the soundness of his work by nature, the infallible judge. He liked also to experiment and to do manual work and nothing pleased him more than combining his own theory with his own experiment.

Fermi was diffident about abstract questions, and he needed absolute clarity in his understanding of them. This may have impeded him at a later date from participating wholeheartedly in the early development of the new quantum mechanics. He did not like philosophy and was very wary of it. In particular, I believe that the philosophical bent of Bohr and Heisenberg did not appeal to him. He could not penetrate Heisenberg's early papers on quantum mechanics, not because of any mathematical difficulties, but because the physical concepts were alien to him and seemed somewhat nebulous. Schrödinger's papers revealed to him the new quantum mechanics; he studied and mastered them as fast and thoroughly as he could, and after he had assimilated the ideas, he believed wholeheartedly in the new developments. He also studied Dirac's papers profoundly, and frequently recast Dirac's ideas in a different mathematical form.

I am anticipating times and developments that were several years off, however. In 1922 relativity was the center of attention, and in this field Italy was up to date, thanks to the pioneer work of Gregorio Ricci-Curbastro and the more recent additions of Tullio Levi-Civita. Thus it is understandable that Fermi's early work was done in this field; he acquired a good reputation as a relativist, and Levi-Civita immediately appreciated his powers. There is a peculiar piece of evidence of Fermi's reputation very early in his career. In 1923, at the height of the relativity fashion, a book on relativity by A. Kopff was translated into Italian and published by Hoepli, a major Italian

Family Background

publishing house. The translator and the publisher had asked several notables, in Italy and abroad, to express their thoughts on relativity, and these essays were published as an appendix to the book. If one is interested in knowing the scientific status of Italy at the time, the essays are instructive; the remarkable thing, however, is that one of the essays was by Fermi, which shows that he was already sufficiently well known to be considered an authority at age twenty-two, barely out of the university.

Fermi's essay "Mass in Relativity Theory" (*FP* 5), among other things, discussed the possible release of nuclear energy and singled it out as the most spectacular consequence of relativity. Fermi quoted Rutherford's experiments on nuclear disintegration, but certainly he had no inkling of things to come or of his own future part in them. It was clear to him, however, that of the two great novelties in physics, relativity and quantum theory, quantum theory had more applications and was likely to give more insight into the constitution of matter. It was pregnant with new physics, rather than mathematics, and hence closer to Fermi's heart.

It was at that time that I first saw Fermi and heard him speak. I was a second-year student at the University of Rome, taking the common preparatory courses for engineering, mathematics, and physics. I had been told by a schoolmate, Giovanni Enriques (son of the mathematician Federigo) that an extraordinary young man named Fermi was to lecture on quantum theory, and I went to the lecture. I had learned some quantum theory on my own (the subject had not even been mentioned in our university courses) and I was very much impressed by the clarity of the presentation, the significance of the discussions, and the speaker's knowledge and self-assurance. I felt that, except for his age, Fermi was the equal of some of the most famous professors of mathematics at the University of Rome for whom we had an almost legendary respect. I do not remember the precise topic of his talk—it could have been an elementary presentation of "The Principles of Quantum Theory" (*FP* 22) —but the favorable impression that I received is still vivid in my mind. Apart from the content of his speech, I remember the deep pitch of his voice and his slow and somewhat singing cadence. When he smiled a little milk tooth he had not shed protruded among his regular adult teeth and gave a peculiar appearance to his smile. I did not talk to Fermi, however, for another few years.

2
APPRENTICESHIP

After he had obtained his doctoral degree, Fermi returned to Rome to stay with his family. He had to find a way of supporting himself, and to him the only desirable positions were within the university. The postdoctoral career of a young scientist in Italy at that time often began by his becoming an assistant to a professor and then obtaining the *libera docenza,* which permitted him to teach certain elective courses at a university. A *libero docente* (free teacher) did not receive a salary; sometimes, however, assistants and *liberi docenti* were asked to teach required courses by *incarico* (assignment) and were then paid for this work. All these functions could be combined in various ways to accommodate the protégés of a professor. Also, a postdoctoral fellowship abroad was possible and desirable.

After a scientist had worked a few years as an assistant and libero docente, a *concorso* (national competition) offered the possibility of a regular chair with tenure. Competitions were held whenever there were vacancies and the faculties felt inclined to fill them. Winning one of these competitions, in which three persons were usually chosen, was the main hurdle of and the key to an academic career. Subsequent promotions came mainly by seniority. An assistant in one of the major universities, upon winning a competition, was usually transferred to a small provincial institution, and further advancement consisted in moves to better universities. The competition was judged by a committee of five professors solely on

Apprenticeship

the basis of scientific papers and other documents that were submitted by the candidates; there were no oral discussions or examinations. Although professorships at any university were nominally equal and equally paid under a civil service system, there was a great difference between being a professor in a major university—for example, Rome, Pisa, Padua, Naples, Bologna, or Turin—and in less important or more remote institutions.

For a scientist, the ultimate distinction was election to the Accademia dei Lincei, the National Academy of Sciences of Italy, which dates back to the time of Galileo. Then as now, the academy published an important part of Italian scientific literature and was internationally known as one of the leading scientific academies of the world. Some professors also became Senators of the Kingdom (members of the Italian upper chamber appointed for life); this was both a political and an academic distinction.

Fermi was fully aware of this system and was eager to reach the top as fast as possible. As an important head start, he already had a reputation, for Puccianti in Pisa and the closely knit group of physicists and mathematicians in the other Italian universities did not ignore the emergence of a brilliant new star. In Rome, Fermi was immediately received into a group of mathematicians that centered on Castelnuovo, Levi-Civita, and Enriques. Fermi also called upon Professor Orso Mario Corbino, then the director of the physics laboratory at the University of Rome and tacitly recognized as the most prominent living Italian physicist. Here, in Fermi's words, is an account of their first meetings.

I first met Senator Orso Mario Corbino when I returned to Rome immediately after my graduation. I was then twenty years old; Corbino was forty-six. He was a Senator of the Kingdom, had been a minister of public instruction, and was universally known as one of the most eminent scholars. Thus it was with understandable hesitation that I introduced myself to him, but the hesitation rapidly disappeared under the impact of his manner—at the same time cordial and interesting—as he began to discuss my studies.

In that period we had almost daily conversations and discussions which not only clarified many of my confused ideas, but aroused in me the deeply felt reverence of the pupil for the master. This reverence steadily increased during the years I was privileged to work in his laboratory.[1]

Professor Orso Mario Corbino,[2] a person who had a great influence on the development of physics in Italy, was born April 30,

Apprenticeship

1876, in Augusta, a small town on the east coast of Sicily. His father, Vincenzo, had studied for the priesthood, but the unrest in Sicily, which culminated in Garibaldi's expedition of 1861 and the annexation of the island to Italy, had diverted Vincenzo from his vocation; he had been drafted into the newly formed national Italian army and served on the mainland for almost seven years. He participated in the campaign of 1866 against Austria, and in 1868 was discharged from the army and returned to Augusta, where he bought a small macaroni factory. The macaroni was handmade by Corbino and his family and sold on the spot.

In 1872 Vincenzo Corbino married Rosaria Imprescia, who came from a relatively well-to-do family of landowners. Rosaria was eleven years younger than the bridegroom—the same age difference as that of Fermi's parents. In her family of many brothers and sisters, property had been left to the male descendants, according to the local custom, and Rosaria had no worldly goods. Although she was very intelligent, she had not been taught to read or write because at that time in Sicily schooling was reserved for males, even in relatively affluent families; it was only in 1908, when Rosaria was over fifty years old, that she taught herself to read and write. Of their seven children—four boys and three girls—the oldest boy, Leone, became a noncommissioned officer in the carabineri, a branch of the army that was assigned to police duty. Orso Mario, the second boy, was the physicist. Lupo emigrated to the United States and died there. The youngest boy, Epicarmo, became a distinguished economist and a professor at the University of Naples, and later a cabinet member in the Italian republic.

Orso, a very precocious child, attended elementary school in Augusta. When he was nine years old he was introduced to the local bishop, who had come to Augusta on a pastoral visit, and the good prelate secured a place in the seminary for Orso in the hope of making him a priest. Young Corbino remained in the seminary until he was eleven; he learned there the rudiments of Latin, but he realized that he did not have a religious calling. At his request, his parents withdrew him from the seminary and entrusted him to a canon in Augusta who prepared him for the final examination of the ginnasio; at that time he was only thirteen, two years younger than the regularly graduating students. He next went to Catania, a larger town about twenty-five miles north of Augusta, where he completed his pre-university education by attending the liceo. He completed the

Apprenticeship

normal three-year course in two years and thus was ready to go to the university about three years earlier than the usual age. During the Catania period, he had also attended physics lectures for older boys that were given by Professor Stracciati, an excellent high school teacher who had done some original work on the thermodynamics of blackbody radiation.

Corbino registered as a student of science at the University of Catania but remained there only one year because a friend induced him to study at Palermo, the capital of Sicily, where he could obtain a better education. At Palermo, Corbino completed the remaining three years of physics under Professor Damiano Macaluso, director of the Physics Institute. Macaluso, a cultivated scientist with a non-provincial outlook, was well-to-do, had traveled abroad, and had met and corresponded with some of the leading physicists of the time, such as H. A. Lorentz. In his youth he had worked in Germany and had written a lucid book on thermodynamics. Macaluso immediately recognized Corbino's exceptional talents, and there were very cordial relations between the master, his family, and the pupil.[3]

After he obtained his doctor's degree—in 1896, when he was only twenty years old—Corbino became for a few months a high school teacher in Catanzaro, a town in Calabria, where his older brother, Leone, was stationed, but soon he was called to the Liceo Vittorio Emanuele in Palermo, where he remained for five years. This was a step forward because now he had access to a laboratory, however modest, and could do some research, although his teaching in the secondary school did not leave him much free time.

In 1898 Corbino discovered, together with Macaluso, the anomalous rotary power of sodium vapor in the vicinity of its absorption lines when it is placed in a magnetic field. This was only two years after the discovery of the Zeeman effect, to which the rotatory power is related. Corbino immediately saw the relation between the two phenomena and gave a satisfactory explanation of the effect. This discovery brought him to the attention of the world of physics, even outside Italy. In later years Corbino published several discussions on magneto-optics and became well known in Italy and abroad for his critical understanding of the subject. He was clearly one of the most promising young Italian physicists and was highly esteemed by Professor Augusto Righi of Bologna, who had worked on electromagnetic waves in the wake of Hertz and was then considered as Italy's leading physicist.[4]

Apprenticeship

After its difficult start, Corbino's career proceeded rapidly. In 1904 he won two competitions for university chairs, one in physics, the other in electrical engineering. He accepted the physics chair and went to Messina, where in 1908 he miraculously escaped an earthquake which killed some two hundred thousand persons. (It was the urge to have direct information about this catastrophe that induced Corbino's mother to learn to read and write.) Shortly before the earthquake Corbino had been called to Rome by the director of the physics laboratory of the university, an old gentleman, Professor Pietro Blaserna, who also was president of the Italian Senate and a friend of the Queen Mother, Margherita. Blaserna had studied under Hermann von Helmholtz, but I do not know that he made any important contributions to physics. He took a great interest in the physics laboratory, however, for which he had planned an excellent building, but what little physics was done was due to the work of other, younger physicists, such as Professor Alfonso Sella. When Sella died unexpectedly, Blaserna decided to bring Corbino to Rome.

Corbino then entered a period of experimental research on his old subject, magneto-optics, and also on the specific heat of metals at very high temperatures and on the Hall effect and related galvanomagnetic effects. He was among the first to recognize the power of the cathode-ray oscilloscope, then called "Braun's tube," and he used it extensively, although it was still a very primitive and exotic apparatus. During the First World War Corbino turned his mind to applied research, first on the dangerous study of the vapor pressure of nitroglycerine and later on the development of power supplies for X-ray tubes.

At the end of the war he became increasingly involved in administrative and political work. His technical proficiency was soon discovered by both the government and the industrial world. The government called him to preside over committees that were charged with the administration of the water resources of the nation, and he soon learned the economic and industrial problems peculiar to Italy. In 1920 the Giolitti government made him a Senator of the Kingdom. In 1921 he received his first cabinet appointment, minister of public instruction, in a Bonomi (moderate Socialist) cabinet. In 1923 Mussolini made him minister of national economics—although he was not, and never became, a member of the Fascist party.

Corbino's first-class scientific mind impressed everybody who came in contact with him, including Fermi and most other physicists of the younger generation. He could orient himself with almost incredible speed toward any problem, scientific or human. Around 1920 he was almost the only member of the older generation of physics professors in Italy who understood and appreciated the recent developments in physics. An excellent speaker, he was endowed with sparkling wit, and his intellectual traits were complemented by a warm and generous personality and a propensity to academic maneuvers. He liked to arrange promotions, transfers, and the like, and usually succeeded in these attempts. He was as shrewd and clever as anybody I have ever met, but what distinguished him from many other equally ardent academic politicians was the loftiness of his purposes and the sureness of his judgment.

I think, however, that despite his great successes Corbino had deep regrets—regrets that his exceptional ability had not carried him as far in science as he might have gone under more favorable circumstances. In 1922, speaking in the Senate as minister of public instruction, he said: "Now, Honorable Senators, I also have passed through a crisis which I want to prevent for my colleagues of tomorrow. For me too there has been a moment of need, small but absolute need. I resisted as long as I could, but ultimately I gave in. I have become a senator, I have become a minister (to the damage of public instruction), but I still yearn for science. I yearn above all, in the bitterness of political action, for the peaceful days passed among experiments and apparatus. And I regret that, after the death of Augusto Righi, Italian physics has been unable to find his successor."[5]

These lines, I think, are the key to much of Corbino's mature life and to his friendly actions with respect to Fermi. I wonder if at the time of this speech he had an inkling that Righi's successor was at hand.

Corbino and Fermi talked not only of physics but also of politics, of world events, and of anything else. This habit, continued over the years, contributed materially to Fermi's education outside of physics. Laura Fermi reports that Corbino commented on Mussolini's 1922 March on Rome as follows:

> On the morning of October 28 Fermi was in Corbino's office. This time they did not talk of physics. Nor did they make plans for future work. Corbino was preoccupied with the political situation. He did not like the profession of violence made by the Fascist leader, Mussolini. That young man was tough, ruthless. The columns that on his order

Apprenticeship

were entering Rome constituted a threat and a danger to the country.

"But," he said, "the cabinet's decree of a state of siege is no solution. It can bring nothing good. If the king signs it, we may have a civil war. The army will be ordered to fight, and if they obey, if they don't pass to the Fascists' side, there is little doubt of the outcome: Facists have no arms; there will be a massacre. What a pity! So many young men will die who were only in search of an ideal to worship and who found none better than fascism."

"You put in doubt the king's signature. Do you think he may go against his cabinet? He has never been known to take the lead but has always followed his ministers."

Corbino pondered a minute. "Yes," he said, "I think there is a chance that the king may not sign the decree. He is a man of courage."

"Then there is still a hope," Fermi said.

"A hope? Of what? Not of salvation. If the king doesn't sign, we are certainly going to have a Fascist dictatorship under Mussolini."[6]

When Fermi related this conversation to his family the same evening, the king had in fact refused to sign the decree and the fascist dictatorship was soon to start according to Corbino's prediction. Fermi considered at that time the possibility of emigrating, but he stayed for sixteen years longer.

Fermi's immediate problem was how to enter a wider, less provincial scientific world. It was clear that he would have to go abroad to acquire firsthand knowledge of current physics and to establish contacts with a circle of physicists more active than those in Italy. The Italian Ministry of Education had one fellowship for postdoctoral study in the natural sciences, and Fermi competed for it. The committee that awarded the fellowship was composed of two physicists, two mathematicians, and one chemist, and its conclusion was unanimous:

Endowed with a powerful mind and having performed deep studies in higher mathematics and in the most difficult questions of physics, he (Fermi) already shows—a few months after having obtained his degrees—a scientific maturity which allows him to treat with clear intuition problems both of mathematical physics and experimental physics. This is demonstrated by his interesting work on the dynamics of a rigid system of electric charges in any configuration and by his work on problems of electrostatics in a uniform gravitational field; by explanation of a fundamental discrepancy between the expressions of the electromagnetic masses; by some papers on pure general relativity pertaining to phenomena occurring in the proximity of a world line; by a study on the center of gravity of electromagnetic masses; and by a note on an important theorem of probability calculus.

Apprenticeship

His papers in experimental physics are also interesting. In them he treats the shape of X-ray lines as determined by the nonperfect parallelism of X-rays, and an application of Gouy's method for obtaining images of the anticathode.

Confronted with such powerful and copious activity at the beginning of a scientific career, we cannot but marvel and formulate the wish that in obtaining this fellowship the candidate will further enlarge the field of his knowledge of physics so as to obtain even greater profit from the studies he has thus far performed.[7]

The fellowship committee met on October 30, 1922; that is, two days after the Fascists' march on Rome and the previously reported conversation between Corbino, who was a member of the committee, and Fermi. The work mentioned in the committee's report comprises the first six papers in the *Collected Papers*. Of these, *FP* 3, "On the Phenomena Occuring in the Proximity of a World Line," has permanent interest; it contains the important result of tensor analysis that Fermi had submitted for publication in January 1922 as mentioned at p. 22.

Fermi used the fellowship in the winter of 1923 and went to Göttingen to work at the institute of Max Born. Activity in physics in Göttingen was then at its height. Born was professor of theoretical physics, James Franck was professor of experimental physics, and around these two men was a group of young people who were destined to change physics. Many of the German theoretical students moved on an almost standard path: to Munich with Sommerfeld, to Göttingen with Born, to Copenhagen with Bohr. When Fermi arrived at Göttingen, he found there several brilliant contemporaries, among them Werner Heisenberg and Pascual Jordan, two of the brightest luminaries of theoretical physics. Indeed the two had already been recognized for their exceptional abilities, and Born was writing papers in collaboration with them at about the time of Fermi's residence in Göttingen.

Unfortunately, it seems that Fermi did not become a member of that extraordinary group or interact with them. I do not know the reason for this. Fermi's German was certainly good enough to allow easy communication. Born was cordial with Fermi, but may not have fully appreciated his ability. Fermi's private papers contain several communications from Professor and Mrs. Born that are couched in very friendly terms, one of which is an invitation to Fermi's sister, Maria, to stay in Göttingen with the Borns. The Borns, moreover,

Apprenticeship

while visiting Rome, called on Fermi's family. The fact is, however, that Fermi's sojourn in Göttingen was not so profitable as might have been expected. Fermi wrote some papers that he could just as well have written in Rome, and he does not seem to have responded to the exciting environment. Is is possible that the physicists of his age group—Heisenberg, Pauli, and Jordan, all exceptional men who should have been Fermi's companions—were so engrossed in their own problems that they failed to recognize his ability? Fermi, moreover, was shy, proud, and accustomed to solitude. Perhaps this is why he remained aloof. Only in later years did he become a good friend of Heisenberg and Pauli.

On May 8, 1924, Fermi's mother died. The blow was not unexpected because she had been ill for some time and was hospitalized in a sanitarium for pulmonary diseases. Because of his reserve and extreme reluctance to show his feelings, we know little of Fermi's relations with his parents. He very seldom mentioned them to his friends, and most of the information I have comes from indirect sources. Undoubtedly Fermi admired his mother. Occasionally he praised her intelligence and ability to his wife, Laura. "If she wanted something, she would make it for herself," Fermi once told his wife, referring to an ingenious homemade pressure cooker devised by his mother. This remark would have applied also to the son, who acquired from her the tendency to do everything by himself, and also his unusual self-reliance. In Enrico's infancy the mother was closer to Giulio and recognized Enrico's talents only in his late childhood, probably because of his less sunny disposition. Giulio's death changed her from a lively to a despondent person, and Enrico found living at home depressing in spite of his affection for his parents and sister.

All the facts indicate that the Fermis were a tightly knit family. From his university years on, whenever free, Enrico returned to live with his parents and sister. In 1924 they acquired a house under construction at the Città Giardino Aniene, on the outskirts of Rome, where they planned to live together. By the time the house was ready, the mother had died; but Enrico, his father, and his sister Maria went to live there. Alberto Fermi died on May 7, 1927, three years almost to the day after his wife's death. Enrico and Maria assisted him in his last illness, taking turns for several weeks staying up at night. After

Apprenticeship

his death they remained together in the same house until Enrico married in 1928, and then Maria stayed on alone.

When Fermi returned from Göttingen, he was given the *incarico* of a course of mathematics for chemists and biologists at the University of Rome, a job traditionally under Corbino's control. Persico was then Corbino's assistant, and thus the two friends could work close to each other. Fermi held this temporary job for the year 1923–24. Then a paper he had written at Göttingen on the ergodic theorem[8] came to the attention of Paul Ehrenfest in Leyden, who was impressed by its content and asked his former pupil George Uhlenbeck, employed in Rome as tutor in the family of the Dutch ambassador, to look up Fermi. Thus began the lifelong friendship between Uhlenbeck and Fermi. Shortly thereafter, on Professor Volterra's recommendation, Fermi obtained a three-month fellowship from the International Education Board, a Rockefeller philanthropy, for study at the University of Leyden.

Before going to Leyden, Fermi spent the summer of 1924 in the Dolomites, hiking and enjoying a simple outdoor life. Fermi would sometimes lie in a meadow, with pencil and paper in hand, and write something he had previously elaborated in his mind. He thus developed a theory of atomic collisions and introduced what is now called the Weizsäcker-Williams method. Its fundamental idea, in Fermi's words, is as follows:

> When an electrically charged particle passes near a given point, a variable electric field is created at this point. If we now decompose this field into harmonic components by Fourier's integral, we see that it is the same as the field which would occur at the point if it were irradiated with light having a suitable continuous spectrum.
>
> Assume now that an atom is located at the point we have been considering. Then it is natural to assume that the electric field of the charged particle produces on the atom the same excitation and ionization processes which would be produced by the electric field of the equivalent light. When we know the absorption coefficient of the atom for light as a function of frequency, we can calculate the probability that a charged electric particle passing with a given velocity at a certain distance from an atom ionizes it.[9]

This method made a deep impression on Fermi, who often used it later. We learn of the manner in which this paper was written from a letter Fermi addressed to Persico asking for the experimental data to be compared with his theory.[10] The paper was published in both

Apprenticeship

Italian and German, which indicates that Fermi thought it was important.

Fermi adhered to a fixed publication policy. For various reasons, including directives of the Fascist government, all scientific work had to be published in Italian; but Fermi knew that very few persons outside Italy read the Italian physics journals. Therefore, whenever he thought he had something interesting to say, he would also publish the paper in German, usually in the *Zeitschrift für Physik,* which was then a leading international journal. In later years he persuaded all of his pupils and friends to adopt the same policy. Fermi was very conscious of the international character of physics, scorned the provincialism of many Italian professors, and strove to enter the international community. On his sojourn abroad he had met a number of leading foreign physicists, and he knew many more through their writings. He correctly judged that the best way for young Italian physicists to gain a reputation outside Italy was by publishing *only* important results that would command the attention of the profession. Papers of minor interest or those that were written primarily to increase the volume of papers to be submitted at competitions were printed only in Italian, and thus remained virtually hidden from the rest of the world. With the advent of the Nazi regime, publication in German was abandoned as a token of protest, and English journals became the favorite outlet for announcing important results.

Fermi's collision paper was sent to the *Zeitschrift für Physik,* and soon after its publication Bohr observed that Fermi's method, when applied to the energy distribution of the ejected electrons, was not borne out by experiment. Bohr criticized the paper: "In these circumstances one cannot consider as support for Fermi's assumptions the fact that an estimate of the stopping power based on the requirement of energy conservation agrees approximately with experiment."[11] Fermi was hurt by this criticism, which he felt was unjustified, and which came at a time when he was struggling for recognition. Probably it had a negative influence on his attitude toward the "spirit of Copenhagen." When quantum mechanics was developed, Fermi's method found its rigorous quantum mechanical justification in the time-dependent perturbation theory, as formulated by Dirac. E. J. Williams and C. F. Weizsäcker carefully discussed its limits of validity, including relativity, and used it in several important applications.

Here we see an early example of Fermi's way of working. A rather

Apprenticeship

limited number of fundamental ideas or methods made a deep impression on him. Among these were the ideas contained in the paper under discussion and, later, the time-dependent perturbation theory and its application to the transition from a state to a continuum. The latter formula he called "golden rule no. 2, a term that has been accepted in common parlance by physicists and has even found its way into textbooks. When confronted with a new problem, Fermi often discovered that it could be envisaged as a simple application of one of his beloved ideas. The connection, which by hindsight might appear natural, was of course by no means easy to find.

On September 1, 1924, Fermi went to Leyden, where he found a very congenial atmosphere. Ehrenfest, who encouraged him greatly, gave Fermi an authoritative judgment on his ability, based on valid comparison, because he was acquainted with almost every theoretical physicist in the world. This was the reassurance the young physicist needed. The problems of interest in Leyden, moreover—statistical mechanics and spectroscopy—were well suited to Fermi. He told me once how he had worked on the brownian motion of a string and solved some apparent difficulties of the problem. It is probable that his persistent reflections on the entropy of a perfect gas, on the Sackur-Tetrode formula, and on Stern's calculation of the entropy constant originated at that time. We find traces of his interest in these problems in several papers that he wrote in 1924 and 1925. The discovery of the future Fermi statistics was thus incubating. It is clear (see for instance *FP* 16, "On Stern's Theory of the Entropy Constant of a Perfect Monoatomic Gas") that Fermi was groping for something that eluded him. We now know this was Pauli's exclusion principle, but falling short of this discovery, which was to be the key to so much of physics, Fermi was unknowingly preparing for its application to statistical mechanics.

On his return from Leyden Fermi looked eagerly for a job that would provide him a livelihood. He was highly esteemed by Puccianti, Corbino, Levi-Civita, and others—all of whom had submitted some of his papers to the Accademia dei Lincei—but Fermi thought he could also advance his career through a substantial number of publications. Thus he wrote (in Italian) as many papers as he could. Although he kept his standards high, it is clear that he counted them carefully and felt satisfaction in seeing the pile of his reprints mount ever higher. He wanted to reach the next step in the academic career, the libera docenza, as rapidly as possible, and he believed that the sheer number of publications was important—especially if the judges should be too lazy, or unable, to assess the value of his contributions.

A new arrangement had to be found by the potentates in Italian physics to provide jobs for Fermi, Persico, Rasetti, and a few others, all of whom were at the same stage in their careers. Assistant positions paid very modestly, although they permitted much freedom and the opportunity for pursuing one's own studies; and very few were available. An agreement nevertheless was reached between professors Puccianti in Pisa, Garbasso in Florence, and Corbino by which Rasetti went to Florence, where he was later joined by Fermi, while Persico remained at Rome.

Fermi took the interim post at Florence at the end of 1924, after his return from Leyden. The physics laboratory of the University of Florence—of recent construction—was located outside the city, at Arcetri, on a magnificent site for a villa but detached from the rest of the university. It had perhaps been built there for the sentimental reason that the aged Galileo had lived close by. The historic site was conducive to eloquent inaugural addresses and other grandiloquent speeches, but these were poor reasons for placing a modern laboratory there. The director was Professor Antonio Garbasso, who also was the mayor of Florence and at that time more concerned with politics than science. Professor Vasco Ronchi conducted work in optics, chiefly the study, testing, and improvement of optical instruments, but Fermi had little connection with this work. Fermi taught theoretical mechanics and electricity at a level that corresponded to a present-day upper division or beginning graduate courses in American universities. He diligently wrote notes for the electricity course, copies of which are still preserved in the files of his former students.

His research was a continuation of some of the work that had been initiated or inspired at Leyden. Most of his papers, however, were based upon his current reading of the literature. Fermi in those years regularly read most of the *Zeitschrift für Physik* and some of the other major journals. He thought deeply about what he read and was often inspired to add something new. This habit, which lasted until the time of his neutron work, helps to explain the vastness and universality of his knowledge.

Apprenticeship

Rasetti's presence in Florence was also important. It gave new impetus—indeed, initiated—experimental work in a way similar to what Fermi's presence was doing for theory. Rasetti's exceptional native ability and versatility made him a precious companion to Fermi, as they had already found out in their student days in Pisa. They could discuss together possible experiments, comment on the scientific literature, and effectively and fruitfully stimulate each other. The overall influence was reciprocal; if Fermi taught theoretical physics to Rasetti, Rasetti taught Fermi many other things ranging from modern English literature to biology, and at the same time Rasetti's exceptional grasp of experimental physics allowed him to do significant, modern experiments with very modest means. The reciprocal influence and scientific collaboration continued uninterrupted until Fermi left Italy.

The most important and successful undertaking in 1925 was the study of the depolarization of resonance radiation under the action of an alternating magnetic field[12] undertaken by the two friends in collaboration. This is the first significant experimental work in which Fermi participated, and it is best described in Rasetti's words:

> Wood and Ellett, and Hanle, had announced their remarkable discovery of the effects of weak magnetic fields on the polarization of the resonance radiation of mercury. Rasetti had observed these effects in Florence. When Fermi came to that university a few months later, he was greatly interested in the phenomenon, whose only theory at the time was a classical one based on the concept of Larmor precession. Fermi pointed out that since the mercury resonance line showed an anomalous Zeeman effect with a Landé factor of 3/2, the mercury atom should more likely precess with a frequency 3/2 times higher than the Larmor frequency. The choice between the two alternatives might be decided by investigating the behavior of the polarization under magnetic fields of the intensity of about one gauss and frequency of a few megacycles per second, in approximate resonance with the precession frequency of the atom.
>
> Rasetti had experience with the spectroscopic technique, but neither of the experimenters had any with radiofrequency circuits. However, Fermi calculated the characteristics of a simple oscillator circuit which should produce fields of the proper strengths and frequencies. Fortunately some triodes were discovered in an instrument cabinet and pronounced by Fermi apt to operate the projected circuit. The laboratory also possessed several hot wire ammeters to measure the current in the coils, in order that the magnetic field strength could be determined. Had these instruments not been available, the experiment could not have been performed, since the research budget of the laboratory was exceed-

ingly meager and did not allow the purchase of costly equipment. Another consequence of this financial situation was the fact that the building was never heated, since it was easy to calculate that one month's heating would have absorbed the entire annual budget of the Physics Department. The temperature in the building from December to March ranged from 3° to 6° C. Unfortunately, in the spring, when the experiments were performed, the room temperature had risen to 12° C, more comfortable for the inhabitants but somewhat too high to ensure a sufficiently low density of the saturated mercury vapor.

Inductance coils and other simple parts were built by the experimenters, and when the circuit was assembled, it instantly worked as Fermi had predicted. The experiments were readily performed; unfortunately the accuracy was poor, due to the high temperature and the photographic method employed for measuring the polarization. Still, the results clearly showed that the precession frequency of the atom agreed with the prediction based on the Landé factor.[13]

This work anticipated by several years the extensive applications of radiofrequency to atomic spectroscopy, to which it is related. There also were attempts, which are mentioned in Fermi's correspondence, to use the electric field of solar light to produce a Stark effect in a variable high frequency electric field, but these attempts were unsuccessful.

At Florence Fermi was preoccupied with his career. He wanted a regular university appointment that would permit him to devote himself to physics, without having to worry about his livelihood. His material needs were few; he lived a Spartan life, and he had no family responsibilities. Thus the modest but assured salary of a beginning professor would have been sufficient for all his material needs. The title of professor would also have given him official recognition, but by now he was well aware of his own ability, and I do not think that vanity or any desire for official recognition played a part in his thinking. A serious reason for his wanting a professorial appointment was his love of teaching, apparent in all his activities from the time of his boyhood. Fermi correctly judged the relative importance of his achievements in physics and his progress in a university career, and he was fully conscious of the overriding priority of the first over the second. He also knew that inevitably his abilities had to be recognized. He was in a hurry, however, and in typical fashion was ready to work hard to obtain his goal. His letters to Persico show the unexpected: Fermi active in academic maneuvers.

That Fermi had been intensely interested in his formal career may come as a surprise to persons who knew him in later years. I remember that as early as 1930, or shortly thereafter, he showed little interests in his friends' and collaborators' problems of academic advancement. He was of course right in giving overriding priority to scientific achievement and to favorable working conditions over career questions; nevertheless, he seems to have forgotten very early the way he himself felt in his youth.

In 1925, although Fermi produced some good original work in physics, career problems predominated. About ten years later Fermi would say that now he knew the rule, "even-numbered years are propitious for physics; odd-numbered years are not."

Fermi spent the summer of 1925 in the Dolomites. As usual, several of the mathematicians from Rome and their families were there to escape the heat of the plains—Levi-Civita, Castelnuovo, Bompiani, Ugo Amaldi, and the younger Francesco Tricomi. R. de L. Kronig, a young brilliant physicist, also joined the company, and he and Fermi went on long hikes with Ugo Amaldi's seventeen-year-old son, Edoardo, who had just finished high school. The boy was fascinated by their conversation, although he understood very little of it. Later, when Kronig left, Fermi and Edoardo Amaldi, the strongest athletes of the company, went off together on a strenuous bicycle tour of the Dolomites.

At last, in the fall of 1925, an opportunity of obtaining a permanent position presented itself. A competition for a chair of mathematical physics at the University of Cagliari in Sardinia was announced. All the competitors who entered it were much older than the two youngest competitors, Fermi and Persico; they belonged scientifically to a different generation. At that time, in Italy, "mathematical physics" was interpreted rather narrowly as elasticity, classical electricity, and so on. Corbino once jokingly said, "It is the theoretical physics of 1830." Modern theoretical physics, in which e and h, the symbols of electron and quantum theory, are mentioned, had no chairs in Italy. Still worse, theoretical physics was viewed with suspicion if not outright hostility by several established professors of physics, who often made fun of it or referred disparagingly to subjects they did not know or understand.

Corbino was the outstanding proponent of theoretical physics among the professors of physics of his generation. He had not mastered all its technical intricacies, but he understood and appre-

Apprenticeship

ciated its main results and problems, recognized its potentialities, and clearly saw the necessity of establishing it firmly in Italy if the country was to attain and preserve a high scientific level. Fermi and Persico, completely committed to the cause, were his competing champions.

The position of the mathematicians was altogether different. Since about 1870 mathematics in Italy had been in a good, and more recently in an excellent, state of development. The mathematicians who were closest to theoretical physics were probably Volterra and Levi-Civita. Volterra was primarily interested in his invention, functionals, and their applications, and Levi-Civita in absolute differential calculus, or tensor analysis (to which he had made fundamental contributions), and its applications to relativity. Both men worked in Rome, were internationally famous, anti-fascists, and preeminently mathematicians rather than physicists. Fermi was closer scientifically and personally to Levi-Civita, but the two men differed widely in interests because of their respective ages, outlooks, and scientific education. Other mathematicians, especially Enriques and Castelnuovo, also had wide interests, but they were even further removed from physics; nevertheless, they were aware of the importance of Fermi's work, and he gravitated socially toward them.

When the committee judged the competition for the chair of mathematical physics, it decided unanimously that, on an absolute scale, Fermi deserved the position, but it split on its comparison of the competitors. Three committeemen (reputedly professors Somigliana, Guglielmo, and Marcolongo) voted for Giovanni Giorgi, the propounder of the MKS system, and two members (reputedly Levi-Civita and Volterra) voted for Fermi. Only one position was available, and Fermi remained at Florence. The report of the committee, published in March, 1926, discussed the work of the participants in the completion and then singled out Fermi and Giorgi.

> More difficult was the choice between Fermi and Giorgi for first place, because both of these excellent competitors are so different in activity and maturity.
>
> Some of the committeemen thought that the greater maturity of Giorgi, his scientific productions, and the character of his mind—speculative and philosophical—formed a complex of elements which made him preferable to Fermi. However, they appreciated the scientific production of the latter and from it derived the most optimistic expectations for the future of this young investigator.

Apprenticeship

Other committeemen, although they concurred with the judgment of their colleagues on the solid and intelligent activity of Giorgi, preferred Fermi because of the importance and originality of his investigations.[14]

Fermi, who thus suffered a setback he felt was unjust, remembered the competition and the judges for many years; from a practical point of view, however, he was hardly damaged, because within one year he was to win a competition for the University of Rome, an incomparably greater success than the Cagliari appointment. In the meantime he had resumed his scientific work.

His earlier meditations on the entropy of the monatomic perfect gas were suddenly illuminated by Pauli's discovery of the exclusion principle. This offered the key to the quantization of the gas through the rule that only one particle per quantum state was allowed. Fermi's paper "On the Quantization of the Perfect Monoatomic Gas" (*FP* 30), dated February, 1926, was presented to the Accademia dei Lincei by Corbino and soon published in longer form in the *Zeitschrift für Physik* (*FP* 31.) This was Fermi's first major contribution to physics.

The problem of the equation of state and of the expression of the entropy of a perfect gas was an old one. Boltzmann and Planck had worked on it, and Sackur and Tetrode, as well as Stern, had found an expression for the entropy of a gas that was consistent with quantum theory. Statistical mechanics, however, encountered difficulties in correctly counting states for gases containing identical particles, but a first important step toward solving these difficulties had been taken by Bose, who derived the blackbody formula by statistical methods in 1924. He treated the blackbody as a perfect gas of light quanta and counted the microscopic states in a novel way. Einstein immediately grasped the importance of Bose's work and extended it to a gas of ordinary molecules. Bose and Einstein's central idea was to consider the system as defined by the number of identical particles occupying the single-particle quantum states. These occupation numbers are all it takes to define the state of the system. This approach differs from the classical one according to which even identical particles are distinguishable inasmuch as they can in principle be followed in their trajectories; this becomes impossible if the de Broglie wave packets, by which particles may be described, overlap considerably. An example will help to clarify the difference between the ways of counting states. Let us assume that we have two identical particles a, b and three individual states 1, 2,

Apprenticeship

3. We put in a parenthesis the number of particles in each of the individual states, for instance (1, 0, 1) means 1 particle in state 1, none in state 2, and 1 in state 3. The possibilities are then:
(2, 0, 0) (0, 2, 0) (0, 0, 2) (1, 1, 0) (1, 0, 1) (0, 1, 1)
i.e., occupation numbers for states 1, 2, 3 are respectively 2, 0, 0; 0, 2, 0, and so forth; of these the first three can be realized in one way only, but the second three are realized each one in two ways according to classical statistics because (a, b, 0), (b, a, 0), and so on, are distinct, while Bose statistics does not distinguish (a, b, 0) from (b, a, 0) and counts them as one. Bose-Einstein statistics applies to all particles of integral spin (photons, pions, He^4, and so forth). Particles of half integral spin (electrons, protons, and so forth), on the other hand, have the peculiarity that only one or zero particles can go into a particular quantum state (Pauli's principle). The permissible occupation numbers are thus only one and zero. Fermi developed the theory for the properties of an ideal gas that is composed of such particles. Today particles of the first kind (integral spin) are called *bosons* and those of the second kind (half integral spin) *fermions*.

In practice, all diluted gases at high temperature behave in the same classical way; only when density is high do peculiar phenomena appear—for instance, the specific heat vanishes. This condition, *degeneracy*, occurs when $nh^3 (3mkT)^{-3/2} >> 1$. Here n is the gas density in particles per unit volume, m the particle mass, T the absolute temperature, and h and k the Planck and Boltzmann constants. To understand the meaning of this relation, one must remember that to localize a particle or wave packet with an accuracy Δx one needs waves of wavelength $\lambda < \Delta x$. Now λ is connected to the momentum by de Broglie's relation, $h/p = \lambda$. Classically, the average kinetic energy of an atom is $p^2/2m = (3/2)kT$, from which we get $\lambda = h(3mkT)^{-1/2}$. To avoid interference phenomena between the waves of different atoms incompatible with the classical description, the respective wave packets must be kept at a distance that is great in comparison with their dimensions. Because the average distance between atoms is n^{-3}, we must have $\lambda << n^{-3}$ or $n\lambda^3 << 1$, which can also be written $nh^3 (3mkT)^{-3/2} << 1$.

It so happens that several very important systems are approximately representable as degenerate gases; for instance, the electrons in a metal, the electrons in an atom, and the nucleons in a nucleus can all, in a first crude approximation, be considered degenerate

Apprenticeship

fermion gases. For this reason Fermi's statistics is very important, and its influence pervades several chapters of physics. Fermi's statistics and its quantum mechanical foundation were found independently by Dirac a few months after Fermi's work.

The applications of Fermi's statistics to metals was almost immediate. For instance, a difficulty of long standing was that the free electrons in a metal did not show any specific heat. This fact was correctly attributed to a degeneracy phenomenon, but the ideas on the subject were not clear. Fermi was well aware of the problem, which had been investigated experimentally by Corbino (among others). Fermi tried to apply his statistics to the heat of vaporization of metals but did not publish anything on the subject. It was Sommerfeld who first applied the new theory to metals in a systematic way. The paper on statistics, universally recognized as a fundamental advance, brought Fermi to the forefront of the international fraternity of theoretical physicists. Fermi now had behind him a new accomplishment that no judge in a competition could overlook. Corbino saw his own plans for fostering modern physics in Italy closer to realization, and he pressed for the establishment of a chair of theoretical physics at the University of Rome. Nor was Fermi idle, as the correspondence with Persico indicates.[15] The competition was to be announced in the fall of 1926.

A more important development for the world of physics, however, and even for Fermi personally, was the appearance of Schrödinger's first papers on wave mechanics. (The publication of these papers extended over the next two years.)[16] Fermi began to study them in Florence and was deeply impressed. He was, naturally, well aware of the problems (or better, the crisis) of quantum theory. He had mastered the old Bohr-Sommerfeld theory; he had read but not appreciated the early papers of Heisenberg on matrix mechanics; but he was ignorant of de Broglie's work when Schrödinger's first paper in the series, "Quantisierung als Eigenwert Problem," hit the world of physics as a thunderbolt. This and the subsequent papers were immediately hailed as a great step forward in quantum theory. Unlike those of Heisenberg and of Dirac, they were easily if not completely understood; they used familiar tools in physics and mathematics, and they gave concrete results subject to immediate experimental verification. The reaction of Planck and the Berlin

Apprenticeship

physicists, vividly portrayed in the correspondence between Planck and Schrödinger, was typical.[17]

Fermi immediately mastered the Schrödinger technique but for some time he was uncertain about the interpretation of the wave function, or *field scalar,* as it was then called. He had read Born's paper on collisions, in which the statistical interpretation had been put forward, and had even made an elegant application[18] of Born's collision theory; but he was still trying different interpretations of ψ several months later.

Meanwhile the competition for the University of Rome was being judged. The committee, which met November 7, 1926, was composed of Garbasso, Maggi, Cantone, Quirino Majorana, and Corbino, who wrote the report. Fermi won first place, Persico second place, and Aldo Pontremoli third place. After discussing the papers submitted by the competitors, the committee commented as follows on Fermi's work:

The committee, having examined Professor Fermi's large and complex scientific work, has unanimously recognized its exceptional qualities and finds that he, even at his young age and after very few years of scientific activity, already highly honors Italian physics.

On the one hand he commands the most subtle mathematical tools; he uses them soberly and with discretion, without losing sight of the physical problems he is trying to solve and the play and magnitude of the physical quantities he is considering. On the other hand he is perfectly familiar with the most delicate concepts of classical mechanics and mathematical physics; he moves with complete assurance in the most difficult questions of modern theoretical physics, in such a way that he is the best-prepared and most worthy person to represent our country in this field of intense scientific activity that ranges the entire world. The committee thus unanimously finds that Professor Fermi highly deserves to have the chair of theoretical physics, the object of this competition, and feels it can put in him the best hopes for the establishment and development of theoretical physics in Italy.[19]

As a consequence of this competition Persico went to Florence and Pontremoli to Milan. Pontremoli perished soon thereafter in a polar expedition by dirigible led by Commander Umberto Nobile. Fermi, having secured the Rome post, had practically attained the zenith of a university career in Italy. His satisfaction was the greater for being reunited with his father and sister.

3
PROFESSOR AT ROME

Fermi, twenty-six years old, arrived in Rome to take an important chair with lifetime tenure—a position that most professors reached only in their fifties. His academic career had reached its summit. There was, however, an important problem still to be solved: he needed an adequate scientific environment.

There was in Rome, as we have seen, a group of eminent mathematicians considerably older than Fermi; he was very friendly with them, but the excellent personal relations and the high opinions they had of each other were not sufficient to create fruitful exchanges or intellectual cooperation. Among the physicists, Corbino, the only one able to follow and appreciate the developments of modern physics, was occupied with political and business activities. Although he was fully aware of the necessity of rejuvenating physics in Italy and of introducing new young workers to the field, his scientific activity was limited. A new generation had to take over, and Fermi was its destined leader.

From his Pisa years Fermi had striven to introduce modern physics in Italy. He considered this one of the main objectives of his life, although he did not say it explicitly and even less did he overtly propagandize for the cause. Grand speeches were alien to his nature, which was more prone to deeds than to talk. To implement his program, therefore, Fermi took several practical steps that were to have important consequences. These were first, to write articles on modern physics intended for a wide audience, including high school teachers; second, to write a textbook devoted to atomic physics; and third, to seek out and train young physicists.

The articles originated in semipopular lectures that Fermi gave on new developments in physics at the yearly meetings of the Italian Society for the Advancement of Science and on other similar occasions. The lectures were often coordinated with speeches by Corbino, Persico, Rasetti, or other proponents of modern physics. The texts of the lectures were then published as articles in the proceedings of the society and reproduced in other technical journals for engineers or teachers. The articles by Fermi, excellent in content, were not always written with stylistic elegance. However, they served the purpose for which they were intended; they helped attract the attention of the younger generation to modern physics.

Until 1928 there was no Italian book on modern physics suitable for training advanced university students. The text from which an entire generation had learned, Sommerfeld's *Atombau und Spektrallinien*, was in German; furthermore, it was too long and too detailed to serve as an introduction to the subject. Typically, Fermi decided that the best thing to do was to write his own book, which he did during the summer vacation of 1927, again in the Dolomites. As before, lying on his stomach in a mountain meadow, armed with an adequate supply of pencils and of bound blank notebooks, he wrote page after page, without a book for consultation, without an erasure (there are no erasers on Italian pencils) or a word crossed out. The handwritten manuscript was then sent to the publisher, Zanichelli of Bologna, and the book was published in 1928. This short book, *Introduzione alla fisica atomica,* performed its task admirably. Unfortunately, it was written before wave mechanics had really taken hold, and for this reason, although Schrödinger's equation is treated in a chapter at the end of the book and Heisenberg's matrices are mentioned, the development of the subject is primarily based on the old Bohr-Sommerfeld orbit conception.

The third step, the discovery and training of young scientists, was the most important and most difficult. To make clear what Fermi accomplished, I shall have to describe the situation in Rome before his arrival.

In 1927 the number of physics students in Italian universities was very small; Rome, one of the largest schools, had perhaps a dozen students in all four years of the curriculum. Physics was taught mainly as a service course for engineers, chemists, doctors, and the like, but was also intended for future high school teachers. Of the two great innovations, relativity and quantum theory, the first was usually mentioned, but in all the Italian universities there was probably no professor except Fermi and Persico who taught quantum theory.

After Fermi moved to Rome, he tried to transform the physics institute (approximately equivalent to an American department) into a first-class institution that would hold its own with the best in the world. He did not express this intention in any document, but it was one of the chief motivations for his actions. On this, he and Corbino were in complete agreement, and thus the two men could collaborate and supplement each other's efforts. One of the first tasks was the strengthening of experimental physics. This proved more difficult than establishing theoretical physics, because there was no experimentalist who could singlehandedly do what Fermi was doing for theory. Furthermore, experimental physics required technical and financial support on a much greater scale than did theoretical work.

Besides Corbino there was at Rome at the time another experimental physicist, Professor Antonino Lo Surdo, who had the chair of *fisica superiore* (advanced physics). Lo Surdo, like Corbino a Sicilian, had worked primarily in spectroscopy and had found an efficient way to observe the splitting and frequency shift of spectral lines emitted by hydrogen in an electric field, a phenomenon commonly called the Stark effect. Lo Surdo's work, which came soon after Stark's discovery, was done around 1914, but after this he did not follow the subsequent developments of physics—at least there are no indications that he did. Lo Surdo's lectures, which I attended in 1927 and 1928, included interesting demonstration experiments, but they were essentially expositions of Drude's book on optics (1900)[1] and J. J. Thomson's work on gas discharges (1903);[2] they made no significant mention of quantum theory. Lo Surdo, however, thought that he represented modern physics and he was jealous of his prerogatives. He had tried to prevent Fermi's call to Rome, feeling that his presence would in some way be a personal slight to him. Lo Surdo's attitude and actions produced strained relations with Corbino, who chose to ignore him within the limits of civility and respect for the prerogatives of another professor in the institute.

Professor at Rome

Lo Surdo, an embittered man who had suffered severe personal misfortunes, could have immensely improved his position by helping the new generation, but he chose to antagonize Fermi and the young physicists and was severely punished by the events. In later years, at the time of the Fascists' racial persecutions, he displayed great loyalty to the party and was duly rewarded for it. He lived, however, to see the collapse of fascism and the waning of the influence he had acquired in the last years of the dictatorship.

The necessity of adding a young, active experimental physicist to the staff in Rome was manifest and urgent, and Rasetti was the obvious choice. He was transferred from Florence early in 1927, receiving the position of first assistant (*aiuto*) to Corbino.

Rasetti's performance was brilliant. In a few years he was to acquire an international reputation, chiefly through his work on the Raman effect, and Corbino arranged the creation in Rome of a new chair of spectroscopy for his protegé. Some years later, in 1935, Corbino said that the neutron work should bring the Nobel Prize to Fermi, election to the Accademia dei Lincei for Rasetti, a physics chair for me, and, to Amaldi, succession to my job as first assistant to Corbino. All this came to pass.

The next step was recruiting students, good students. Those Fermi found at the institute were not of the quality he desired, and the problem of selective recruitment was not easy. In Italy at that time, physics opened only the possibility of an academic career. The university chairs were very few, and generally it took a long time to reach one of them. High schools offered only very modest positions. Thus a student of physics with ambition for success had to have a true passion for his work or independent means, or a combination of both. However, Fermi's reputation had spread to circles outside the physics institute, and I for one, from the engineering school, felt as soon as Fermi arrived that a great opportunity was at hand.

In the spring of 1927 I met Rasetti through my friend and schoolmate, Giovanni Enriques, the son of the mathematician. While mountain climbing with Rasetti, I learned several physical theories from him; but even more important, I gained the conviction that physics in Rome was awakening. I had already heard Fermi speak at the mathematics seminar of the university in 1924, and he had impressed me deeply. I had never been taught atomic physics in my regular university studies but had caught glimpses of it while reading Reiche's book on quantum theory.[3] Now I encountered in Rasetti and Fermi two young men deeply versed in the subject. I had still not met Fermi personally, however, and I had no way of comparing the young lecturer of a few years earlier with the famous physicists I knew through my readings.

During the summer of 1927, through Rasetti and Enriques, I came to know Fermi and immediately had the impression of having found an exceptional teacher. Fermi, in conversations on hikes and at the seashore, asked me several simple questions in mathematics and classical physics, perhaps to test discreetly my knowledge and ability. In September 1927, on my return from an expedition to the Alps with Rasetti, Enriques, and others, I went to the International Physics Conference in Como and there had the unforgettable experience of seeing in person the great physicists whose names I had read in books—Lorentz, Rutherford, Planck, Bohr, and many others —and a group of very young men—Heisenberg, Pauli, and Fermi— obviously destined to be the "great" of the future. I also learned a considerable amount of physics at the conference because Rasetti and Fermi would point out the various celebrities and tell me about their most important achievements.

It was clear that, scientifically at least, Fermi's was the most significant voice from Italy at the conference.[4] He was one of the very few Italians whose work was mentioned and who participated in the discussions. Famous physicists such as Born, Sommerfeld, and others listened carefully to what he had to say and obviously considered him one of the important physicists at the conference.

When I returned to Rome in the fall of 1927, I decided to heed my old desire to study physics, and a few months later I left the engineering studies that I had been pursuing and transferred to the physics institute. On that occasion I discovered Corbino's far-reaching influence. He easily overcame serious administrative difficulties so as to give me credit for my engineering studies. No one took special notice of it, but I had become Fermi's first pupil, at least in the formal sense, and the Roman school had started.

In June 1927 Corbino announced to one of his engineering physics classes that the physics institute was ready to accept a few new students of uncommon ability. He pointed out that there were exceptional opportunities because of the scientific promise of modern physics and the arrival of new young faculty members versed in it and eager to teach. He stressed the requirements of the prospective recruits. Only one student, Edoardo Amaldi, accepted the invitation and became a member of the small group of students who were to be personally tutored by Fermi in the hope of developing research physicists and collaborators. Eventually this might lead to a renaissance of physics in Italy, but the goal was not explicitly formulated at the time.

After a few months of study I talked to my friend and schoolmate at the engineering school, Ettore Majorana (1908–38),[5] who also joined our group. He was a nephew of Professor Quirino Majorana, an experimental physicist at the University of Bologna, one of the judges in the competition that brought Fermi to Rome. In intellectual power, depth and extent of knowledge, Ettore Majorana greatly surpassed his new companions, and in some respects—for instance, as a pure mathematician—he was superior even to Fermi. Unfortunately his exceedingly original and profound intelligence was accompanied by a tendency to exaggerated criticism and by a deep pessimism. His nature led him to work alone and to live a solitary life. He did not participate much in our studies but he helped us with theoretical difficulties that came up and stunned us with his original ideas and methods and with his ability as a lightning-fast mental calculator (he could easily have been a professional numerical prodigy). Later he isolated himself even more; by 1935 he had disappeared from the university and seldom left his home.

Other students frequented the institute and attended Fermi's regular lectures on electricity, modern physics, etc. Occasionally they came to Fermi's private seminar; among them Gabriello M. Giannini, who emigrated to the United States in the late twenties and in due course became a successful industrialist and businessman.

Fermi's seminar was always improvised and informal. In the late afternoon we would meet in his office, and our conversations might give rise to a lecture—for example, if we asked what was known about capillarity, Fermi would improvise a beautiful lecture on its theory. One had the impression that he had been studying capillarity up to that moment and had carefully prepared the lecture. I find in one of my notebooks on the discussions of those years the following topics: blackbody theory, viscosity of gases, wave mechanics (the establishment of Schrödinger's equation), tensor analysis, optical dispersion theory, gaussian error curve, more quantum mechanics, and Dirac's theory of the spin.

In this fashion we reviewed many subjects at a level that corresponded to a beginning graduate course in an American university.

Professor at Rome

Sometimes, however, discussion was on a higher level, and Fermi might explain a paper he had just read. In this way we became conversant with some of the famous papers by Schrödinger and Dirac as they appeared. We never had a regular course. If there was an entire field of which we knew nothing and about which we asked Fermi, he would limit himself to mentioning a good book to read. Thus, when I asked for some instruction in thermodynamics, he told me to read the book by Planck. But the readings he suggested were not always the best, perhaps because he mentioned only the books he had studied himself, which were not necessarily the best pedagogically but simply those he had found in the library at Pisa. After his lecture we would write our notes on it and solve (or try hard to solve) the problems he had given us, or others we thought of. We spent the rest of our time on experimental work. The instruction was chiefly in theoretical physics, but no distinction was made between future theoreticians and experimentalists. Fermi, who at the time worked mainly in theory, also was interested in experimental work. I vividly remember one of my first visits to the laboratory: Fermi and Rasetti—one short and stocky, the other tall and lean, both in not-too-clean, gray smocks—focusing the fringes of a Jamin interferometer on the slit of a Hilger quartz spectograph encased in shiny mahogany. They tried to awe me with their work, the measurement of the refractive index of thallium vapor. The results—or results very similar to them—were published in a paper[6] and stemmed directly, at least in technique, from Rasetti's doctoral thesis.

Rasetti's influence on Fermi and on the whole group was great, even outside physics. He read books (fiction and popular science), he traveled to remote places, he collected insects, he ate special foods, and so on. By subtly extolling his own readings or activities he spurred imitation. We called him the "revered master" (*venerato maestro*) in a joking way which had more than a grain of truth in it.

A comic reflection on the community of life and intensity of involvement of the group was that everyone adopted a peculiar voice and accent. Both Fermi and Rasetti had developed a deep voice and a slow, strangely modulated cadence that inadvertently was copied by all of their friends. It is said that one of them, while traveling by railroad, struck up a conversation with another passenger and was very surprised when he was asked if he was a physicist in Rome. He inquired how his companion had guessed this, and the reply was, "from your way of speaking."

Professor at Rome

The seat of all this activity was the old physics laboratory of the University of Rome in Via Panisperna 89A. The building was erected about 1880 on what had been the periphery of the city and on land of a monastic order that had been secularized about 1870. (The neighboring chemistry department even used some buildings of the old monastery.) The physics building was perfectly adequate for scientific work in the 1920s, and it compared favorably with other major European laboratories. The equipment was fair, including mainly instruments for optical spectroscopy and some adequate subsidiary apparatus. The shop was old-fashioned and the machine tools rather poor, but the library was excellent. The location of the building in a small park on a hill near the central part of Rome was convenient and beautiful at the same time. The garden, landscaped with palm trees and bamboo thickets, with its prevailing silence (except at dusk, when gatherings of sparrows populated the greenery), made the institute a most peaceful and attractive center of study. I believe that everybody who ever worked there kept an affectionate regard for the old place, and had poetic feelings about it.

The third floor of the building was Corbino's residence. The second floor contained the research laboratories and the offices of Corbino, Lo Surdo, and Fermi, as well as the library. The first floor contained the shop, the classrooms, and the students' laboratories. The basement contained the electric generators and other facilities.

Fermi, Rasetti, and their students occupied the whole south side of the second floor; Lo Surdo most of its north side. As time went on, the number of persons associated with Fermi increased so greatly that they occupied most of the second floor. Neighboring quarters were occupied by Professor G. C. Trabacchi, who was the chief physicist at the Health Department (Sanità Pubblica). He had an excellent supply of instruments and materials which he generously lent whenever we needed them, a fact that earned him the nickname of "divine providence." Corbino used to spend part of the morning in his study and often visited Fermi's quarters and stopped to talk about physics and other subjects with him and his students.

The first important paper Fermi wrote after his appointment to Rome was "A Statistical Method for the Determination of Some Atomic Properties" (*FP* 43): the Thomas-Fermi statistical method. Fermi did not know that Thomas had invented the method almost a year earlier. With his usual vigor he calculated the key function

$\varphi(x)$ on a small hand calculator. Majorana, who had just joined the institute, being a very skeptical man, decided that he had better check the numerical work on the $\varphi(x)$. He went home, transformed the equation for φ into a Riccati equation, and solved it numerically without the help of a calculating machine of any sort. When he compared his result with Fermi's and found that Fermi had made no errors, he could not hide his surprise.

Fermi then proceeded with many specific applications and suggested several of them to Rasetti and to his students. The fundamental idea of this statistical method became one of his favorites, and he turned to it many times in later years. It is possible that he had it in mind when he made an application of statistical methods to high-energy physics in the paper "High Energy Nuclear Events" (*FP* 241). He summarized his 1927–28 results for a meeting in Leipzig in 1928, sponsored by Debye; this was one of the first (perhaps the first) international meeting in which Fermi had a leading part. His participation in the Sixth Solvay Congress of October 20–26, 1930, however, was an even more important recognition than Leipzig. The Solvay Conferences had been established by Ernest Solvay, the Belgian inventor of a process for preparing sodium carbonate. This successful invention had made him rich, and he devoted part of his fortune to endow the Conferences. The meetings were held in Brussels and lasted about one week, during which about thirty of the most prominent and active physicists of the world discussed, privately, one specific subject. The first meeting had been held in 1911, and it was on "Radiation and Quantum Theory." The sixth, the first in which Fermi participated, in 1930, was on Magnetism. The only other Italian ever invited to these select working conferences was Corbino, in 1924, but he had not gone.[7]

Fermi's knowledge and interests embraced all of physics, and he diligently read several journals. As has been noted, he preferred concrete problems and distrusted theories that were too abstract or general, but any specific problem in any field of physics—classical mechanics, spectroscopy, thermodynamics, solid-state theory, and several more—fascinated him and constituted a challenge to his ingenuity. Often, when just talking with him, one heard a beautiful explanation develop, simple and clear, which would resolve a puzzling phenomenon. We repeatedly had occasion to witness in this way the execution of a new and original piece of work. What one saw was the development of a calculation at a moderate speed but with exceedingly few errors, false starts, or changes of direction. The work proceeded almost as in a lecture, although more slowly, and at the end, the paper or at least the equations were ready to be copied for publication with little need of improvement. It is impossible, of course, to say how much preliminary work Fermi had done, consciously or unconsciously.

A notable example of this mode of operation was "Quantum Theory of Radiation" (*FP* 50). Fermi had seen the papers by Dirac[8] and understood the results obtained. Dirac's method, however, was alien to Fermi, and he decided to recast the theory in a form that was mathematically more familiar. Almost every day he would show Rasetti, Giulio Racah, Majorana, Amaldi, and me what he had done, and often he would make a new application in our presence. He did not read, at least at that time, the important papers by Jordan, Pauli, and Heisenberg on quantum electrodynamics. He tried very hard—filling many notebooks with calculations—to find an escape from the divergencies problems of quantum electrodynamics; but he was not successful. Fermi described his work on quantum theory of radiation in a series of lectures he gave at the Institut Poincaré in Paris in 1929, and in more complete form at the summer school in Ann Arbor, Michigan, in 1930. The lectures were published in French and in English; and Bethe, who collaborated with Fermi on a problem of quantum electrodynamics, described their impact:

> Many of you probably, like myself, have learned their first field theory from Fermi's wonderful article in the *Reviews of Modern Physics* of 1932. It is an example of simplicity in a difficult field which I think is unsurpassed. It came after a number of quite complicated papers on the subject, and before another set of quite complicated papers on the subject, and without Fermi's enlightening simplicity I think many of us would never have been able to follow into the depths of field theory. I think I'm one of them.[9]

The comments of Wigner, who did not need to learn the subject from Fermi, are equally illuminating:

> Fermi disliked complicated theories and avoided them as much as possible. Although he was one of the founders of quantum electrodynamics, he resisted using this theory as long as possible. His article on the Quantum Theory of Radiation in the *Reviews of Modern Physics* (1932) is a model of many of his addresses and lectures: nobody not fully familiar with the intricacies of the theory could have written it, nobody could have better avoided those intricacies. However, when he tackled

Professor at Rome

a problem which could not be solved without the explicit use of the much disliked concepts of quantum field theories, he accepted this fact and one of his most brilliant papers is based on quantized fields.[10]

Wigner is hinting here about destruction and creation operators; it is likely, however, that Fermi did not fully understand "those intricacies" until 1933, shortly before the work on beta decay.

A peculiar characteristic of Fermi's working habits was the steady pace at which he proceeded. If the passages were easy, he nevertheless proceeded quite slowly, and a simple-minded observer might have asked why he wasted so much time on simple algebra. However, when difficulties arose that would have stopped a man of lesser ability, Fermi solved them without a change of speed. One had the impression that Fermi was a steamroller that moved slowly but knew no obstacles. The final result was always clear, and often one was tempted to ask why it had not been found long ago inasmuch as everything was so simple and natural. Once used, a method was stored in his memory and often was adapted to problems that appeared to be very different from the one that had originated the physical idea or mathematical technique. The evolution and successive applications of the "scattering length" in "On the Pressure Shift of Lines High in Spectral Series" (*FP*95) and "Artificial Radioactivity Produced by Neutron Bombardment" (*FP*107) and the recurrence of the statistical theme applied to atoms and nuclei are examples of Fermi's adaptive technique.

From as early as 1928 Fermi made little use of books; Laska's collection of mathematical formulas and Landolt Börnstein's tables of physical constants were almost the only reference books he had in his office. When he needed a complicated equation from a book in the library, Fermi would often propose a wager, saying he would derive the equation before we could find it in a book; and usually he won. The only treatise that I know he read after he came to Rome was Weyl's *Gruppentheorie und Quantenmechanik*.

The speed at which it was possible to train a young physicist at "Fermi's school" was incredible. Naturally a good deal of the success was due to the immense enthusiasm that had been aroused in the young people—never by exhortations or "sermons" but by the eloquence of example. After having spent time in the institute in Via Panisperna, one became completely absorbed in physics, and in saying "completely" I am not exaggerating.

Fermi did not like to assign subjects for doctoral dissertations or, in general, to suggest subjects of investigation. He expected the students to find subjects for themselves or to obtain one from a colleague who was more advanced in his studies. The reason for this, as he later told me, was that he did not easily find subjects simple enough for beginners; he usually thought of problems that interested him personally but were too difficult for students. He treated his students and collaborators in a friendly and direct manner that was free of complications—never unkind but never overconsiderate. He would overcome the key difficulty in an investigation by one of his young associates and generously refuse even mention of his help, but he demanded maximum effort. He inspired great respect. Nobody would have dared mention to him a petty squabble or jealousy such as may occur in a collaborating group. Rasetti was very generous in teaching the experimental techniques he knew and in lending apparatus he no longer used, but it was difficult to work with him because he had peculiar idiosyncrasies and irregular schedules of work.

Strong and long-lasting friendships developed among all the participants in this adventure. Age differences were small: Fermi and Rasetti, the oldest, were twenty-six years old in 1927; Amaldi, the youngest, was nineteen. Corbino seldom came to the afternoon lectures, but he took a strong interest in the scientific work and in the welfare of the group, including the career problems and relations of the young men with the outside world.

Another revival of physics, in many ways similar to the one occurring in Rome, also was in the making in Florence. Persico had a strong and beneficial influence on this movement, being the only tenured professor of the younger generation. Of the able young scientists, all in their twenties, who congregated in Florence and formed a group of physicists working on modern problems, the most famous are Bruno Rossi, of cosmic ray fame; Giuseppe Occhialini, the co-discoverer of the showers and of the pion; Gilberto Bernardini, later the director of physics at CERN (the European Center for Nuclear Studies in Geneva) and director of the Scuola Normale at Pisa; and Giulio Racah, inventor of the Racah coefficients and later the rector of the University of Jerusalem. The Florentine physicists, mostly self-taught, were full of vigor, enthusiasm, and original ideas. The groups in Florence and Rome enjoyed excellent relations and frequently exchanged visits and seminars. It would take me too long and lead too far afield, however, to detail the distinguished history of the Florentine group, except for its direct connections with Fermi.

Word of what was happening in Rome and Florence spread quickly among young Italian physicists or aspiring physicists. An early and frequent visitor in Rome was Giovanni Gentile, Jr., who had recently graduated at the Scuola Normale at Pisa. In 1930 Persico moved from Florence to Turin, where he helped establish another center of modern physics. Renato Einaudi was the first to come to Rome from Turin on Persico's recommendation and was followed by Gian Carlo Wick. Wick had obtained his doctor's degree at Turin and had worked with Born at Göttingen and Heisenberg at Leipzig, where he met Amaldi. Amaldi tried to arrange Wick's move to Rome, but a lack of available positions retarded the project one year. In 1932 Wick came to Rome where he remained until he was appointed professor at the University of Palermo at my instigation. Later, several university students came to Rome as soon as they had finished their two years of preparatory courses: Eugenio Fubini and Ugo Fano from Turin, Leo Pincherle from Bologna, and Bruno Pontecorvo from Pisa. These transfer students represented a younger group, almost a second generation.

By 1929 it was clear that, whereas the theoretical situation was well in hand, it was necessary to strengthen our experimental activities. In order to import new experimental techniques to Rome, members of the group had to work in different laboratories to learn them on the spot. Thus Rasetti went to Pasadena, to Millikan's laboratory, where he did important work on the Raman effect; I went to Amsterdam, to Zeeman's laboratory, to study forbidden spectral lines; and Amaldi went to Debye's laboratory in Leipzig, where he worked on X-ray diffraction of liquids. Initially we used laboratory facilities that were not available in Rome for the completion of work we had already started at home; later, we used foreign laboratories for experimenting in entirely new fields. In this second phase, Rasetti worked on radioactivity in Lise Meitner's laboratory at the Kaiser Wilhelm Institut in Berlin-Dahlem, and I worked on molecular beams in Otto Stern's laboratory in Hamburg. The plan was successful, for without these periods of foreign training in experimental research, it would have been impossible later to perform the complex neutron work rapidly and efficiently. Even while we were abroad, letters kept us in close contact with the group in Rome, and during vacations we discussed theoretical problems with each other. There are many traces of this exchange in the papers by Fermi, Majorana, Wick, and others. During this period, Fermi went abroad only for short visits. He was by now accustomed to being somewhat isolated intellectually because only Majorana (who was rather inaccessible anyway) could speak with him about theory on an equal footing. On the other hand, in the very early 1930s a number of young physicists, attracted by the rising fame of Rome and/or worried by the threatening German political situation, began to arrive in Rome. Hans Bethe, George Placzek, Felix Bloch, Rudolph Peierls, Lothar Nordheim, Fritz London, Edward Teller, Christian Møller, Sam Goudsmit, and others spent some time in Rome, often on their way to the United States. From among the Americans, we had a visit by Eugene Feenberg, whom Majorana especially liked; their mutual attraction manifested itself in their sitting in the library facing each other in silence because they knew no common language.

The longer visits and the presence of these men added greatly to the scientific life of Rome. We had the opportunity of seeing some representatives of German science at its best, and they in turn learned some of Fermi's peculiar strengths. It was in Rome, for example, that Bethe wrote his famous article "One- and Two-Electron Systems" for the *Handbuch der Physik*. I remember that he stood at a high table, and without books wrote page after page of this treatise without interruptions, and recalculated all results of previous authors to check their correctness.

Bethe reported his impressions of Fermi as follows:

My greatest impression of Fermi's method in theoretical physics was of its simplicity. He was able to analyze into its essentials every problem, however complicated it seemed to be. He stripped it of mathematical complications and of unnecessary formalism. In this way, often in half an hour or less, he could solve the essential physical problem involved. Of course there was not yet a mathematically complete solution, but when you left Fermi after one of these discussions, it was clear how the mathematical solution should proceed.

This method was particularly impressive to me because I had come from the school of Sommerfeld in Munich who proceeded in all his work by complete mathematical solution. Having grown up in Sommerfeld's school, I thought that the method to follow was to set up the differential equation for the problem (usually the Schrödinger equation), to use your mathematical skill in finding a solution as accurate and elegant as possible, and then to discuss this solution. In the discussion, finally, you would find out the qualitative features of the solution, and hence understand the physics of the problem. Sommerfeld's way was a good one for many problems where the fundamental physics was already understood, but it was extremely laborious. It would usually take several months before you knew the answer to the question.

It was extremely impressive to see that Fermi did not need all this labor. The physics became clear by an analysis of the essentials, and a few order-of-magnitude estimates. His approach was pragmatic. . . .

Fermi was a good mathematician. Whenever it was required, he was able to do elaborate mathematics; however, he first wanted to make sure that this was worth doing. He was a master at achieving important results with a minimum of effort and mathematical apparatus.

By working in this manner he clarified the problems very much, especially for younger people who did not have his great knowledge. For instance, his formulation of quantum electrodynamics is so much simpler than the original one of Heisenberg and Pauli that it could be very easily understood. I was very much intimidated by the Heisenberg-Pauli article, and could not see the forest because of the trees. Fermi's formulation showed the forest. The same was true in the paper we wrote together, concerning the various formulations of relativistic collision theory. Fermi's formulation of neutron diffusion, the age theory, has been exceedingly fruitful in making quick calculations of neutron diffusion even in complicated cases. I could multiply this list easily, just from my own experience with Fermi and his work.[11]

Special mention should also be made of George Placzek, who came for longer periods than the others, worked experimentally in Rome, learned Italian well, and became a close personal friend of several members of our group. He was concerned—as many of us were—with the Raman effect, especially of ammonia; and working with Amaldi, he obtained good Raman spectra of this gas with techniques that had been developed by Rasetti. The Raman spectra of gases were difficult to obtain and required long exposures with expensive quartz lamps, which had a limited life (I remember his dismay when a very valuable plate fell from his hands and broke). Like Bethe, Placzek wrote a classic article in Rome, "Rayleigh Streuung und Raman Effekt," for the *Handbuch der Radiologie,* in which he developed his new ideas on molecular symmetry. Placzek had wider interests than most physicists; he knew many languages and literatures and had deep and wide historical and political knowledge. A man full of wisdom, wit, and integrity, he contributed to everybody's education even beyond the field of physics, and his untimely death in 1955 was sorely felt.[12]

Life in Via Panisperna was very methodical. One worked from

about 9:00 A.M. to 12:30 P.M. and from 4:00 P.M. to 8:00 P.M. Of course, this schedule was self-imposed and everybody was free to keep it as he wished. Work in the evening after dinner was practically unknown, and on Sundays we often went for a hike in the vicinity of Rome, or for a mountain trip. During the winter there were skiing expeditions; and during the summer, trips abroad or vacations in the Alps. The most significant personal event of this period was Fermi's marriage to Laura Capon, on July 19, 1928. Laura Fermi's *Atoms in the Family* contains a vivid account of her meeting with Fermi, their marriage, and their family life. They had two children: Nella, born January 31, 1931; and Giulio, born February 16, 1936.

In 1929 Mussolini appointed Fermi to the newly created Accademia d'Italia. The appointment, which was made without an election, had an interesting backstage history. In 1928 Corbino had gone to the United States on business and thus was absent from Rome when the 1928 elections to the Accademia dei Lincei took place. He wanted to have Fermi elected to the Academy, and he had left his letter nominating Fermi with his colleague, Lo Surdo, and had asked him to read it at the proper time. When Corbino returned, he found that Fermi had not been elected; and when he questioned Lo Surdo, the latter answered that in his absent-mindedness he had forgotten Corbino's letter. There was nothing Corbino could do at the time, but I can imagine that he was deeply offended and desirous of revenge. An occasion soon presented itself.

Mussolini distrusted the Accademia dei Lincei because he thought (quite correctly) that many of its members—at least of its influential and distinguished members—were hostile to fascism. He had limited himself to ignoring it, however. He now decided to form the Accademia d'Italia, which would outdo the Lincei. He could not give his creation the distinction of age, but he could give its members an additional substantial salary (more than one received from a university chair), a picturesque uniform, the title of "Excellency," and high precedence in official ceremonies—things the Lincei did not have. In his appointments he chose as much as possible men who combined professional distinction and friendliness to fascism. Fermi was the only physicist he appointed. The honor, though deserved, was unexpected because Fermi's reputation still was limited to physics and, according to tradition, at his age academic honors were not yet due. Furthermore, Fermi had no political standing with the Fascist party; in fact, he could be best described as nonpolitical. Other persons—notably Lo Surdo—had high hopes of receiving the honor. It is probable that Fermi's appointment was contrived by Corbino, who being a senator was ineligible, and who, although not a Fascist, had been a member of one of Mussolini's early cabinets.

Fermi was very much pleased by the honor, though he was, and remained, singularly devoid of vanity, a weakness common in the academic world. In later years, when he had accumulated an impressive number of honorary degrees, medals, and memberships in academies, he replied to an official questionnaire requesting a list of the organizations to which he belonged: "I am a member of a number of learned societies, the names of which I have forgotten." The Accademia d'Italia appointment, at any rate, had a beneficial effect on the subsequent development of physics in Italy because the physics representative in the academy was undoubtedly the most highly qualified man available, and he became influential even without trying. However, even after his appointment to the Academy, Fermi wielded relatively little political influence because he was unwilling to sacrifice his time to occupations other than physics and he did not like to participate in administrative or political affairs.

Membership in the Academy changed Fermi's financial condition. He was able to resign his posts on the *Enciclopedia Italiana Treccani* and the National Research Council, which he had accepted on his arrival in Rome, mainly to improve his financial position. As in everything he undertook, Fermi had been most conscientious in his work for the *Enciclopedia*. He had given the job the required time, had performed all the editorial chores expected of him, and had written some articles of which he was justifiably proud. His article on statistical mechanics is still an excellent presentation of the subject and has been reproduced in the *Collected Papers* (*FP* 83). At the National Research Council he had examined inventions and research proposals, and had done this thoroughly; but as soon as it was possible, he relinquished the time-consuming job. He passed his jobs on to younger colleagues, who welcomed a way of earning money at a time when it was difficult for a physicist to find gainful employment outside the very few university appointments that were available.

The appointment of Fermi to the Academy and the great fanfare with which it was established gave us hope that the government would seriously help science, and we persuaded Fermi to use his prestige to effect this end. He made a few perfunctory attempts to increase research facilities by the creation of new jobs and subsidies, but these efforts went no further than obtaining Mussolini's approval in a memorandum that was later pigeonholed. The all important university appointments remained under the control of the old guard, and it was only years later that the younger generation could prevail in this important matter.

Fermi visited the United States for the first time during the summer of 1930. Rasetti had been in Pasadena in 1929 and on his return to Rome had filled our heads with the wonders of California. I still remember his descriptions of the orange groves near Pasadena, of his winter ascent of Mount Whitney, of the wonderful work that was being done at the California Institute of Technology and of the pretty girls of Berkeley. It is possible that Rasetti's enthusiasm influenced Fermi. At any rate, when the University of Michigan invited him to teach theoretical physics at a summer session, he accepted happily. He found two old friends, Uhlenbeck and Goudsmit, the discoverers of the electron spin, who had moved from their native Holland to Ann Arbor at the instigation of Professor Walter Colby, who was on the lookout for talent to build up theoretical physics at his university. Ehrenfest joined the group for the summer, and the atmosphere was most congenial. Fermi lectured on the quantum theory of radiation (*FP* 67) and explained that rather new and difficult subject very beautifully.

The visit to Ann Arbor was a great success scientifically and was a pleasant interlude for Fermi—so much so that he became one of the most frequent attendants of the summer school there, returning in 1933 and 1935. Through these visits he came to like America and to appreciate the opportunities it offered. He was attracted by the well-equipped laboratories, the eagerness he sensed in the new generation of American physicists, and the cordial reception he enjoyed in academic circles. Mechanical proficiency and practical gadgets in America counterbalanced to an extent the beauty of Italy. American political life and political ideals were immeasurably superior to fascism. These considerations prepared him to emigrate, and when the time came, the decision was more the fulfillment of a long nurtured plan than an emergency escape.

I accompanied him to America in 1933 but found myself less

Professor at Rome

resistant to the humidity and heat of the summer. If I could not work as hard as Fermi, however, I tried to emulate his swimming across cool lakes. We had excellent dinners in the country, where we learned to appreciate rural American cooking, and finally bought a car, "The Flying Tortoise," which we drove back to New York, not without some mechanical difficulties along the way. These did not scare Fermi, who was a good mechanic. Once in a gas station he showed such expertise in repairing an automobile that the owner instantly offered him a job—and these were depression days.

The most important scientific event of those years, the formation of quantum mechanics, had taken place without major contributions from Italy, at least in the establishment of the principles, although Fermi had contributed to its applications. He had developed his statistics independently of quantum mechanics and before mastering the new theory. In his original papers we can trace the effort to clarify and assimilate quantum mechanics between 1926 and 1931.[13] Schrödinger's papers, previously mentioned—the first he understood—had aroused his great enthusiasm, and Fermi explained them immediately to his friends and later to Corbino, who remained skeptical for some time. Fermi read Dirac's great papers as soon as they appeared in the *Proceedings of the Royal Society* and meditated on them deeply. Later he spoke on the subject to the mathematics seminar, where the professional mathematicians, older and less familiar with the experimental background of physics, raised several ingenious objections to the commonly accepted interpretation of quantum mechanics. Thus "The Interpretation of Causality in Quantum Mechanics" (*FP* 59) originated from a discussion in 1930 in which Professor Castelnuovo raised many questions. Fermi was inclined to be a little impatient with people who did not understand the new developments of quantum mechanics, but he treated genuine difficulties, like those Castelnuovo raised, very differently from the way he treated silly objections, of which there were many.

Fermi complained now and then that even persons for whom he had the highest respect and admiration, such as Corbino, occasionally were skeptical about quantum mechanics and its interpretation, a skepticism he attributed to a lack of understanding. Resistance to quantum theory was shown mainly by persons older than Fermi; the young physicists either understood, or believed, the new theories and in any case learned to use them even if they had not

64

Professor at Rome

completely assimilated them.[14] It must be said, however, that in his last years Fermi seemed less convinced that the current interpretation of quantum mechanics was the final word on the subject.

The advent of quantum mechanics, in the opinion of Fermi and Rasetti and also of Corbino, signaled the completion of atomic physics. Once the fundamental questions relating to atoms were solved, the future lay in the exploration of the nucleus or of complicated structures leading ultimately to biology.

These ideas suggested a radical change in the research projects of the institute because our experimental tradition in Rome went back to the spectroscopic work initiated by Rasetti under Puccianti at Pisa. All our successes in experimental physics up to that time had been on spectroscopic subjects; our equipment was spectroscopic; and our knowledge was mainly in the field of atomic physics. The work of Rutherford and his school was rather alien to us. Thus the change to nuclear subjects cost us considerable effort. It was not a whim or a desire to follow a fashion but the result of a deliberate plan that Fermi and his friends debated vigorously, even heatedly.

Corbino voiced these ideas in an eloquent speech delivered September 21, 1929, at the meeting of the Italian Association for the Advancement of Science at Florence.[15] In this speech, titled "The New Goals of Experimental Physics," he classified past and present investigations of physicists in three categories: (1) discovery of entirely new phenomena, such as the electric current in Volta's cell, X-rays, radioactivity, etc.; (2) verification and application of current theories; and (3) precise measurement of material and universal constants. He then tried to predict the development of these three aspects of physics. He considered them in reversed order, starting with precision measurements. He pointed out that these are mostly performed in specialized laboratories specially equipped. He emphasized their interest and the importance of using pure, well-defined materials such as single crystals of metals. He stressed the interest of the universal constants.

Speaking of verification of theoretical predictions, he pointed out the advanced state—or more accurately, the completeness—of quantum mechanics and the absolute necessity of mastering it for all future investigations. This part of his speech, slightly polemical, emphasized the distinction between classical mathematical physics, as widely practiced in Italy, and the new theoretical physics, exemplified by Fermi's work (which he mentioned specifically). He then discussed the interaction between theory and experiment in fields where existing theory, except for mathematical complications, should be adequate for making valid predictions. He gave criteria to evaluate the importance of studies on which there are theoretical predictions, and discussed the Raman effect and the Davisson Germer experiments as examples. He then said:

One field of study where theory is lagging behind experiment is the clarification of the molecular or atomic ordering in solids and liquids. It has been ascertained that the cohesive forces are of electrical origin, and X-ray analysis has demonstrated the location of groups of atoms in the crystal lattice. However, theoretical prediction of the physical constant of atomic-electronic ensembles is just beginning, and thus there will be much for theoretical physics to do in this field. This study, also, is far from being exhausted from the experimental point of view. . . .

Physics of solid states and liquids and the effects of high pressure or of very low or very high temperatures thus must be considered a field that is still full of promise for theoretical and experimental physicists of today and of the future, in addition to having very important applications.

He finally comes to the first category: entirely new discoveries. After some remarks which tended to discourage the search for new phenomena on the basis of random guesses, a mode of research sometimes practiced in Italy during that period, he said:

Thus the only possibility of great new discoveries in physics is offered by the chance that someone will succeed in modifying the atomic nucleus. This will be the really worthy goal of physics of the future.

Corbino then proceeded to show that particle bombardments were the only way to achieve this goal:

The only way that remains is artificially to produce projectiles similar to those of radioactive bodies but in much greater quantity and with greater velocity. This requires discharge tubes with a potential difference above 10 MeV.

Only technical and financial difficulties, not insurmountable in principle, oppose the realization of this great project. The object is not only the artificial transmutation of elements in appreciable quantities but the study of the tremendously energetic phenomena that would occur in some cases of disintegration or recombinations of atomic nuclei.

Note that, as we have said above, the nucleus is formed by protons or hydrogen nuclei and electrons. However, in the combination, for instance, of four protons to form the helium nucleus, the mass of the compound is somewhat smaller than the mass of the four protons that have combined.[16] This decrease in weight, which is called mass defect, must according to the theory of relativity be accompanied by the emission of enormous quantities of energy. Thus in the formation of the helium nucleus from four protons, for each gram of helium formed one should obtain about 1½ billion large calories, corresponding to approximately 2 million kilowatt-hours of energy. Of course, the inverse phenomenon, namely, the decomposition of a gram of helium into hydrogen, would require the application of the same amount of energy. In these phenomena of nuclear physics, whose immense importance is obvious, one would effect the transformation of matter into energy, and vice versa, at the rate of 25 million kilowatt-hours for each gram of matter transformed.

Thus we can conclude that although it is improbable that experimental physics will make great progress in its ordinary domain, there are many possibilities in the attack upon the atomic nucleus. This is the true field for the physics of tomorrow. To participate in the general movement, however, either in the present directions or in the direction of a more remote future, it is indispensable that experimentalists acquire a ready and sure grasp of the results that theoretical physics is reaching and that at the same time they acquire ever greater experimental means. To try to work in experimental physics without an up-to-date knowledge of the results of theoretical physics and without huge laboratory facilities is the same as trying to win a modern battle without airplanes and without artillery.

Corbino then discussed what might happen if nuclear physics, contrary to his hopes, proved rather barren. In such case, he said, one would have to find new subjects rather than spend money and brains on an effort that looked unpromising. He then concluded his speech by saying:

Thus, even if physics should tend toward a saturated condition, the study of its application to other disciplines—biology for instance—provided it is entrusted to real experts who know all the resources of modern physics, could lead to results of the greatest scientific and practical value. It would be even better if it were possible to combine in one brain the mental attitude of the biologist and the mental attitude created by the new physics, rather than the simple superposition of techniques.

Corbino's speech, followed the same day by an address by Fermi, "The Experimental Foundations of the New Physical Theories" (*FP* 56), and another by Persico, "The Causality Principle in Modern Physics," was an attempt to call the attention of the Italian scientific public to the revolution that had occurred in physics. The speeches indeed made an impression and were widely discussed among scientists, but the reactions were not always favorable. A distinguished critic was Professor Garbasso, who took special issue with

Professor at Rome

Corbino,[17] and a friendly dispute followed in several technical journals. In the last sentence of his last reply to Garbasso, Corbino summarized the issues:

> Having circumscribed the fields of endeavor of future research; having dissuaded those who would undertake a sterile chase after some great new phenomenon; having advocated a more serious study of modern theory to avoid useless investigations; and last, having called attention to the speed and power with which others are attacking the problems of the hour, I cannot be reproached for damaging the progress of the study of physics in our country.[18]

The first step toward implementing these ideas and changing the institute's experimental orientation was Rasetti's stay for several months in Lise Meitner's institute at Berlin-Dahlem to learn nuclear techniques. Work on the hyperfine structure of spectral lines, due to the nuclear spin, a suitable transitional subject between atomic spectroscopy and nuclear physics, occupied Fermi in 1930 and 1931, and we switched our reading to matters connected with the nucleus. Amaldi initiated a systematic study of Rutherford, Chadwick, and Ellis's "Radiations from Radioactive Substances" in a special seminar that was attended by Fermi, Rasetti, Majorana, myself, and occasionally by students. Amaldi and Fermi also started building a cloud chamber as an exercise for acquiring practical skills. However, the old atomic and molecular subjects were still actively pursued.

A nuclear physics conference organized by the Accademia d'Italia from October 11 to 18, 1931, helped to familiarize us with current nuclear problems.[19] It was attended by about forty-five leading physicists, but it came at the wrong time—just before the marvelous discoveries of the neutron, deuterium and positron, which changed so much of nuclear physics.

In July 1932 another nuclear conference was held in Paris as part of a large international conference on electricity. Fermi, asked to report on the status of physics of the nucleus, gave a review[20] in which among other things he stressed the difficulties of a nuclear model based on electrons and protons as constituents, and mentioned Pauli's hypothesis of the neutrino.

During the preceding winter the experiments on the bombardment of beryllium and other light elements with alpha particles, initiated by Walther Bothe and H. Becker around 1930 and pursued by Irène Curie and her husband Frédéric Joliot with a cloud chamber, had given results not easy to explain. The experimenters found a very penetrating radiation and at first believed it was a gamma radiation; but when these "gamma rays" impinged on hydrogen, they projected protons of high energy. The interpretation tentatively suggested by the Joliots that this phenomenon was a Compton effect on protons was hardly tenable, and early in 1932 James Chadwick at the Cavendish Laboratory demonstrated beyond doubt that the radiation contained neutral particles of a mass close to that of the proton, which he called neutrons. Chadwick's experiments demonstrating the neutron were published after Fermi's report had been submitted to the conference. While there was still uncertainty about the Bothe and Curie Joliot results, Ettore Majorana, at Rome, had grasped the significance of the proton recoils observed by the Joliots and, with characteristic irony, commented that they had discovered the "neutral proton" and had not recognized it.

Majorana then proceeded to develop a nuclear model based on neutrons and protons, without electrons, to analyze the forces between neutrons and protons in considerable detail,[21] and to calculate the binding energies of several light nuclei. When he told Fermi and some of his friends of this work, its importance was immediately recognized, and Fermi urged Majorana to publish it, but Majorana demurred because he thought his results, thus far, were too incomplete. Fermi then asked for permission to report Majorana's results at the Paris conference, giving Majorana proper credit for the new ideas, but Majorana said he would give permission only if his ideas were attributed to an old professor of electrical engineering expected to be present at the conference. This strange proposal, showing Majorana's whimsicality, was obviously unacceptable, and so Majorana's ideas were promulgated only much later, after they had been independently discovered by other physicists.

In Fermi's report one can discern uneasiness about the electron-proton model and the current confusion about the neutron. In the discussion that followed Fermi's communication, according to the published minutes of the conference, there was an interesting exchange with the Polish physicist Wertenstein. The minutes read:

> Mr. Wertenstein asked for an explanation of the possibility of the emission of radiation accompanying the beta rays of naturally radioactive bodies and reestablishing the conservation of energy. He did not believe that these rays could be neutrons because of their mass. Mr. Fermi answered that such neutrons are not those that have been discovered, but that they would have a much lighter mass.[22]

Professor at Rome

This exchange shows that the neutrino hypothesis was already known by physicists who were interested in beta decay. In fact, Pauli had been advocating the neutrino since 1930, to try to solve the fundamental difficulties presented by beta radioactivity. This type of decay seemed to violate the conservation of energy because a nucleus starting from a definite state landed in another definite state, but the electron emitted had variable energy. Many experiments had sought to detect the missing energy—for instance, in the form of gamma rays —but to no avail. In order to salvage the conservation of energy and other fundamental conservation laws which seemed violated in beta decay, Pauli postulated a particle, the neutrino, which would take away the energy but would be practically undetectable. He had even asked experimentalists to try to detect the neutrino in a letter addressed to the meeting of physicists at the University of Tübingen in December 1930.[23] In 1931, during the Rome conference, there had been discussions on the neutrino between Pauli and Fermi.

Indeed, it was at the Rome conference that the term *neutrino* had entered the world of physics. Before that time there had been some confusion between a hypothetical neutron, which had not yet been discovered, and the particle that would reestablish the conservation of energy and momentum in beta decay. The term *neutrino* had been suggested by Fermi in informal conversations and represented an application of the endings *one* and *ino* that can be appended to Italian nouns to denote bigness and smallness. The Italian word for neutron, *neutrone,* suggests a large neutral object, and *neutrino* suggests a small neutral object. The term was soon adopted at the University of Rome and from there propagated to the world physics community.

The 1932 Paris conference was followed by the Seventh Solvay Congress in Brussels on October 22–29, 1933. Blackett, Bohr, Bothe, Chadwick, Cockcroft, Marie Curie, Louis de Broglie, Dirac, Ellis, Fermi, Gamow, Heisenberg, F. Perrin, F. Joliot, I. Curie-Joliot, E. O. Lawrence, Meitner, Pauli, Rutherford, and others attended this meeting. By this time the neutron had been discovered and the model of the nucleus, composed of protons and neutrons, had been established through the work of Heisenberg, Ivanenko, and Majorana. Also, the cyclotron had started to perform properly; deuterium was becoming a relatively common substance; and the positron had been established. Seldom had a branch of science been in such a period

Professor at Rome

of bloom as nuclear physics at that time. The meeting was historic both because of the subjects discussed and because of the persons present.

Although in Brussels Fermi's role was confined to comments on problems of nuclear forces, he absorbed many of the ideas that were ventilated. Fermi heard again of the neutrino hypothesis for beta decay, and when he returned to Italy he must have continued to think about Pauli's idea, because only two months after the Solvay conference he wrote his fundamental paper on beta decay.

When Pauli had first made his suggestion in 1930, it was believed that electrons existed in the nucleus, and Pauli thought of the hypothetical neutral particle as another nuclear constituent of small but finite rest mass. Fermi had mastered the problem of electromagnetic emission through the application of Dirac's theory of radiation, but he had some difficulty with the creation and destruction operators of Dirac-Jordan-Klein-Wigner, and he tried to avoid them in his first papers on the quantum theory of radiation. Later, however, he thoroughly understood them; and once he had that technique well in hand, he considered beta decay theory a good exercise in the use of creation and destruction operators.

Thus Fermi wrote his famous paper on the explanation of beta decay inspired by an analogy to electromagnetic radiation emission. Fermi's work transformed the qualitative hypothesis of Pauli into a quantitative detailed theory of great predictive value. By introducing a new type of force, the "weak interaction" described by a proper hamiltonian, he could calculate the relation between energy of the decay and mean life in beta decay, the energy distribution in the electron spectrum, and many other things. A new fundamental constant, now called the Fermi constant, plays for the weak interaction a role analogous to that of the charge of the electron in electromagnetism, and Fermi determined it from the available experimental data. The paper contained all the basic ideas that dominate the field of the weak interactions, and had fundamental importance for the future development of nuclear and particle physics. The idea of the neutrino and the relation between particles and field quanta, for example, proved important in inspiring Yukawa's theory of nuclear forces.[24] Fermi's paper, written at the end of 1933, has stood the test of time with singular success; in fact, except for the nonconservation of parity, even today very few changes would have to be made to it. To be sure, it is incomplete, but the fundamental ideas are still valid.

Professor at Rome

The eminent physicist Wigner, whose style is rather learned and very different from Fermi's, comments:

> The paper is pervaded with an apparent naïveté which invites criticism and generalizations and a more learned presentation. In this writer's opinion this apparent naïveté is characteristic of Fermi's taste and did not represent his state of knowledge when he wrote the beta decay article. He certainly could have added to it even at that time a good deal of abstract material which others would have considered highly significant.[25]

This is certainly true, but Fermi strove always for simplicity, and his uncanny choice of the vector interaction was correct. Thus, rather than discussing many possibilities, either by insight or by luck he gave the correct one.

Fermi gave the first account of this theory to several of his Roman friends while we were spending the Christmas vacation of 1933 in the Alps. It was in the evening after a full day of skiing; we were all sitting on one bed in a hotelroom, and I could hardly keep still in that position, bruised as I was after several falls on icy snow. Fermi was fully aware of the importance of his accomplishment and said that he thought he would be remembered for this paper, his best so far. He sent a letter to *Nature* advancing his theory, but the editor refused it because he thought it contained speculations that were too remote from physical reality; and instead the paper ("Tentative Theory of Beta Rays" [*FP* 76]) was published in Italian and in the *Zeitschrift für Physik*. Fermi never published anything else on this subject, although in 1950 he calculated matrix elements for beta decay as an application of the nuclear shell model.

This work was barely finished when *Comptes Rendus* and *Nature*[26] carried the stunning announcement that Curie and Joliot, by bombarding boron or aluminum with alpha particles, had obtained new radioactive isotopes of nitrogen and phosphorus respectively, which emitted positive electrons—that is, positrons. It was the discovery of artificial radioactivity, which opened entirely new prospects in nuclear physics.

This discovery gave us the occasion to initiate really important new experimental work. In the previous two years Rasetti had learned how to prepare neutron sources by evaporating polonium on beryllium. Some apparatus, including a cloud chamber, had been built, partly in the shop of the laboratory, but mostly by outside contractors; and some Geiger-Müller counters were in operation. All this had been financed by the Italian National Research Council at a cost of two to three thousand dollars a year, a very substantial sum in Italy at that time. While preparing these techniques, Fermi and Rasetti had also tested them—for instance, in a gamma ray spectrograph—but little of importance had been accomplished in the preparatory phase. After the discovery of Curie and Joliot, it occurred to Fermi to use neutrons for producing artificial radioactivity instead of alpha particles, which are repelled by the nuclear charge of the target nucleus. The weakness of the available neutron sources would be compensated by the greater cross section for producing nuclear reactions that could be expected for neutrons.

In March 1934 Fermi suggested to Rasetti that, using neutrons as bombarding particles, they try to observe effects similar to those seen by Curie and Joliot; and near the end of the month several elements were irradiated with Rasetti's polonium-plus-beryllium neutron source and tested for activity. The results were negative because of the weakness of the source. Rasetti then left for a vacation in Morocco and Fermi continued the experiments. He had the idea, essential for success, of replacing the polonium-plus-beryllium source with a much stronger radon-plus-beryllium source. Radon could be employed because beta and gamma radiations would not interfere with the observation of a delayed effect. Professor G. C. Trabacchi had a radon plant and gave the material to Fermi. His very generous cooperation was essential for the success of the work on neutrons that followed during the next two or three years. Radon-plus-beryllium sources were prepared by filling a small glass bulb with beryllium powder, evacuating the air, and replacing the air with radon. The sources decayed with the half-life of radon, 3.82 days. When Fermi had his stronger neutron source, he systematically bombarded the elements in the order of increasing atomic number, starting with hydrogen and following with lithium, beryllium, boron, carbon, nitrogen and oxygen, all with negative results. Finally, he was successful in obtaining a few counts on his Geiger-Müller counter when he tried fluorine. The sources he used were about 50 millicuries of radon-plus-beryllium and thus were quite weak. The Geiger-Müller counters were primitive and the detection apparatus anything but elegant, but the devices worked adequately and were used to the utmost. His first positive results were announced in a letter to *Ricerca Scientifica* on March 25, 1934.[27] The "I" indicates that Fermi expected a long series of communications on the same subject. In fact he published ten.

Fermi, who wanted to proceed with the work as quickly as possible, asked Amaldi and me to help him with the experiments. Rasetti, still vacationing in Morocco, was notified by cable and was asked to come back as soon as possible and participate in the work. Meanwhile Fermi, Amaldi, and I did our best without him. We organized our activities in this way: Fermi did a good part of the measurements and calculations; Amaldi took care of what we now call electronics; and I secured the substances to be irradiated, the sources, and the necessary equipment. This division of labor, however, was by no means rigid, and each of us participated in all phases of the undertaking, proceeding at very great speed. It soon became apparent that chemical separations would be essential for the investigation, and we searched for a professional chemist who could help us. We were lucky: Oscar D'Agostino had been a chemist in the laboratory of Professor Trabacchi and at the time we began our neutron work held a fellowship in Paris at the laboratory of Madame Curie, where he was learning radiochemistry. When he returned to Rome for the Easter vacation he visited Trabacchi, who introduced him to us. We showed him what we were doing and Fermi asked him to join us. We barely knew each other at the time, but soon we became excellent friends and D'Agostino never used his return ticket to Paris. Soon Rasetti returned from Morocco and joined in the work.

The importance of the work on artificial radioactivity produced by neutron bombardment was obvious to us and to all nuclear physicists. To communicate it rapidly to our colleagues, we wrote almost weekly short letters to *Ricerca Scientifica,* the journal of the National Research Council, and obtained what we would now call preprints of these letters. The preprints were then mailed to a list of about forty of the most prominent and active nuclear physicists all over the world, and the letters appeared a couple of weeks later in the journal. It was only one month after the beginning of this work that Lord Rutherford wrote a letter in long hand to Fermi:

23rd April, 1934

Dear Fermi,

I have to thank you for your kindness in sending me an account of your recent experiments in causing temporary radioactivity in a number of elements by means of neutrons. Your results are of great interest, and no doubt later we shall be able to obtain more information as to the actual mechanism of such transformations. It is by no means clear that in all cases the process is as simple as appears to be the case in the observations of the Joliots.

I congratulate you on your successful escape from the sphere of theoretical physics! You seem to have struck a good line to start with. You may be interested to hear that Professor Dirac also is doing some experiments. This seems to be a good augury for the future of theoretical physics!

Congratulations and best wishes,

Yours sincerely,
Rutherford

It is easy to follow the chronological development of the work during the spring and early summer of 1934 by reading the *Ricerca Scientifica* letters. Our first steps were the obvious ones: irradiate all the substances we could lay our hands on. Luckily, we obtained a small subsidy from the Italian National Research Council, approximately 20,000 lire (about $1,000 at the time), which we could spend with complete freedom. I became treasurer and carried this sum in my pocket when I visited various shops to find the substances to be irradiated. We paid cash and no red tape was involved, with the result that we had extremely prompt service and no overhead. The irradiation of elements proceeded methodically, and the capture of neutrons followed by emission of protons or alphas—(n, p) and (n, α) reactions—were soon identified by chemical analysis. We found out also that sometimes the radioactive substance was isotopic with the target. We thought that the more energetic the neutrons the greater their effectiveness in producing reactions; how wrong we were we would discover only six months later.

Proceeding according to increasing atomic number, we finally irradiated thorium and uranium. Because the activities we produced were weak, before irradiation, it was necessary to purify uranium of all beta-active products that would mask any artificial radioactivity we had produced. This was a tedious task and allowed us to observe the artificial radioactivity only for a short time, because the natural growth of beta activity in purified uranium, after a while, prevented us from detecting the artificial activity. We thought that the irradiation of uranium should produce transuranic elements, for which we expected properties similar to those of rhenium, osmium, iridium, and platinum. This erroneous expectation was then common. Only four or five years later it was shown by workers in various countries, independently of each other, that the transuranic elements would not behave like Re, Os, Ir, and Pt but would form a second family of rare earths.[28] We proceeded to show that uranium irradiated with neutrons did not produce any elements with atomic numbers between those of lead and uranium; the proof for this was obtained and was experimentally correct. The possibility of fission, however, escaped us, although it was called specifically to our attention by Ida Noddack,[29] who sent us an article in which she clearly indicated the possibility of interpreting the results as splitting of the heavy atom into two approximately equal parts. The reason for our blindness is not clear. Fermi said, many years later, that the available data on mass defect at that time were misleading and seemed to preclude the possibility of fission. At any rate we believed we had produced transuranic elements when the intense work of 1934 came to a temporary halt with the summer vacation.

Traditionally, the academic year was closed by a solemn convocation of the Accademia dei Lincei, attended by the king of Italy, at which a member of the academy gave a speech on a subject of his special competence. Corbino was the speaker in 1934, and in his address, "Results and Perspectives of Modern Physics,"[30] he described the neutron work performed in his institute and touched upon the transuranic elements.

> The case of uranium, atomic number 92, is particularly interesting. It seems that, having absorbed the neutron, it converts rapidly by emission of an electron into the element one place higher in the periodic system, that is, into a new element having the atomic number 93. The new element also is radioactive and undergoes further disintegrations which are not yet well established. The periodic system of Mendeleev predicts for the new element, because of the position it occupies, some chemical properties similar to those of manganese and rhenium. In fact, the chemical reactions by which it is separated from other substances agree with the anticipated behavior. Obviously, further tests are necessary and many of them have been performed, all with favorable results. However, the investigation is so delicate that it justifies Fermi's prudent reserve and a continuation of the experiments before the announcement of the discovery. For what my own opinion on this matter is worth, and I have followed the investigations daily, I believe that the production of this new element is certain.

This pronouncement, which had not been cleared with Fermi, was seized by the Italian and the international press and given great publicity. Fermi was dismayed—I seldom saw him in such a dark mood

Professor at Rome

—and Laura Fermi tells us that he could not sleep at night.[31] To claim a discovery that was not absolutely certain went so much against Fermi's grain that the well-intentioned enthusiasm of Corbino caused him great anxiety. He was not angry with Corbino, but thought he had acted rashly. Fermi discussed the problem openly with him, and the two of them did everything possible to quiet the ferment in the press. The question of transuranic elements nevertheless remained a cause of uneasiness for Fermi, who was suspicious of the interpretation of the phenomena and strongly resisted the temptation to give new names to the so-called transuranic elements. His most sanguine view of the interpretation was expressed in his Nobel speech, when the mystery was about to be cleared by the discovery of fission.

The neutron work, accomplished by the summer of 1934, was summarized in a paper that was communicated by Lord Rutherford to the Royal Society of London. The manuscript had been prepared in Rome and was delivered by Amaldi and me, who visited Cambridge during the summer. Lord Rutherford immediately read the manuscript with great attention, made several corrections to improve our English, and turned the manuscript over to the Royal Society. When I asked him if it would be possible to obtain speedy publication, he answered: "What do you think I was the president of the Royal Society for?" Unfortunately, my understanding of Rutherford's English at the time was imperfect and I could not follow some of his remarks, which must have been humorous because he laughed with great glee.

One of the questions which seemed of great importance to us was whether the reactions that made an isotope of the target were (n, γ) or $(n, 2n)$, and we tried to solve this problem by cross-bombardments. T. Bjerge and H. C. Westcott were working at Cambridge on similar problems, and Amaldi and I joined forces with them. We found what we thought was a clear example of an (n, γ) reaction, which was an important and correct finding, but we were wrong in some of the steps by which we reached the correct conclusion.

In the summer of 1934 Fermi went to South America for a lecture tour sponsored by the Italian government. In São Paulo he found the Italian physicist Gleb Wataghin and an old schoolmate of Pisa, the mathematician Luigi Fantappiè, who were there as professors on long-term assignments under the joint sponsorship of Italy and Brazil. Wataghin and Occhialini, who had joined him later, remained in Brazil until after World War II and were instrumental in establishing an important Brazilian physics school. Fermi lectured, in Italian, to overflowing audiences in Brazil, Uruguay, and Argentina and was pleased and surprised by the interest in his work shown by the public. On the homebound boat he met the famous musician Ottorino Respighi. They became friends and had long conversations, although it was difficult for them to exchange ideas in their different fields of accomplishment: the physicist was eager to learn the theory of music, but he wanted it explained in terms of physics, while the musician saw it from his professional, artistic point of view. Fermi developed a high regard for Respighi and often mentioned him in later years.

On his return to Europe, Fermi stopped in London to attend an international physics conference, and reported on his work with neutrons.

Earlier, Amaldi and I, having returned from Cambridge, tried to strengthen our conclusions on the (n, α) versus $(n, 2n)$ reactions by irradiating other substances besides those used by Bjerge and Westcott. We thought we had found another clear example of the (n, γ) reaction in aluminum and immediately communicated this result to Fermi, who was still in London. He mentioned our experiment at one of the meetings. Shortly after this I caught a cold and could not go to the laboratory for several days, but Amaldi repeated our experiments and found a different decay period for irradiated aluminum, which indicated that our so-called (n, γ) reaction had not occurred. This was hurriedly relayed to Fermi, who was angry and embarrassed at having communicated an erroneous result. On his return to Rome he scolded us for our apparent carelessness. We were not only unhappy, but confused, because we could find no fault with the various experiments that gave contradictory results.

Within a few weeks the mystery was further deepened by other inconsistencies, and then it was solved. Before this happened, a new recruit was added to our group: Bruno Pontecorvo, who came from Pisa, his home town, and had recently finished his doctoral dissertation in Rome. Pontecorvo had been a close family friend of Rasetti, and he had transferred to Rome to continue his studies under him.

Professor at Rome

We now asked Pontecorvo to help us in our work, and from then on he participated in all our research.

It was now the beginning of the school year 1934–35 and we resumed our irradiation experiments, again in a systematic fashion. In the summarizing paper published in *Proceedings of the Royal Society*,[32] the activity that had been induced in various elements was classified qualitatively as strong (s), medium (m), or weak (w). The classification seemed unsatisfactory, however, and we decided to establish a quantitative scale of activability, even if it was purely arbitrary. Amaldi and Pontecorvo, who were assigned the study of this problem, began to search for conditions of irradiation that would give easily reproducible results, using the 2.3-minute period induced in silver as a standard. They immediately encountered difficulties because the intensity of the radioactivity that was obtained depended unpredictably on the conditions of irradiation. The wooden table that supported a spectrograph seemed to have miraculous properties: when silver was irradiated on that table, it became much more active than when it was irradiated on a marble shelf in the same room. The explanation for this, we found later, was that the neutron source on the wood gave slow neutrons whereas on the marble it did not.

To solve the mystery we made systematic observations, and the recording of data (now in the archives of the Domus Galilaeana in Pisa) began October 18, 1934. The first step was a series of measurements taken by Amaldi inside and outside a small lead housing *(castelletto)* whose walls were 5 cm thick. The measurements showed that outside the housing the activation decreased greatly with increasing distance from the source, but that this did not happen inside the housing.

The next day similar measurements were made in an effort to explain the action of the surrounding lead, and particular attention was focused on absorption and scattering measurements in lead. In the next step, a lead wedge was prepared for insertion between the neutron source and the detector with the idea of comparing its attenuating effect with that of a wide brick of lead. This is a standard method for comparing scattering with absorption cross sections. The wedge was ready a few days later, but Fermi suddenly decided to try filters of light elements first.

Fermi commented on his decision many years later in the course of a conversation with astrophysicist Subrahmanyan Chandrasekhar

Professor at Rome

on the process of discovery in physics. According to Chandrasekhar, Fermi said:

> I will tell you how I came to make the discovery which I suppose is the most important one I have made. We were working very hard on the neutron-induced radioactivity and the results we were obtaining made no sense. One day, as I came to the laboratory, it occurred to me that I should examine the effect of placing a piece of lead before the incident neutrons. Instead of my usual custom, I took great pains to have the piece of lead precisely machined. I was clearly dissatisfied with something; I tried every excuse to postpone putting the piece of lead in its place. When finally, with some reluctance, I was going to put it in its place, I said to myself: "No, I do not want this piece of lead here; what I want is a piece of paraffin." It was just like that with no advance warning, no conscious prior reasoning. I immediately took some odd piece of paraffin and placed it where the piece of lead was to have been.[33]

It is very hard for me, after so many years, to remember exactly what happened, but there is no doubt that paraffin was tried first on the morning of October 22. The experiment was performed during an examination period, and several of us were in another part of the building; but Persico and Bruno Rossi were kibitzing on our work. About noon everybody was summoned to watch the miraculous effects of the filtration by paraffin. At first I thought a counter had gone wrong, because such strong activities had not appeared before, but it was immediately demonstrated that the strong activation resulted from filtering by the paraffin of the radiation that produced the radioactivity. We tried a few other substances as filters and found the powerful effects occurred only with paraffin. By the time we went home for lunch and our usual siesta, we were still extremely puzzled by our observations. When we came back at about three in the afternoon, Fermi had found the explanation of the strange behavior of filtered neutrons. He hypothesized that neutrons could be slowed down by elastic collisions and in this way become more effective—an idea that was contrary to our expectations. The same afternoon each of us thought of, and repeated, some of the recent troublesome experiments with the slowing down hypothesis in mind. In about half an hour we had the explanation of the disagreement between the results of the different irradiations in aluminum: slow neutrons produced one activity by the (n, γ) reaction; fast neutrons another by the $(n, 2n)$ reaction. This was a great relief to Amaldi and me because it explained a serious puzzle and showed that there was no need to correct the minutes of the London conference. Fermi went to the extreme of hypothesizing that the neutrons could be thermalized (that is reach the energy corresponding to thermal agitation, about 0.03eV), and that same day he devised an experiment (unsuccessful at the time) to test this hypothesis by slowing down the neutrons in a hot medium rather than a cold one and observe some difference in their behavior.

That same evening, at Amaldi's home, we prepared a short letter to *Ricerca Scientifica* (*FP* 105). Fermi dictated while I wrote. He stood by me; Rasetti, Amaldi, and Pontecorvo paced the room excitedly, all making comments at the same time. The din was such that when we left, the maid shyly asked Amaldi's wife Ginestra whether her guests were tipsy. (The Amaldis had been married about a year earlier.) Ginestra Amaldi, then working with *Ricerca Scientifica*, handed the manuscript to the director of that journal the following morning.

The translation of the letter follows:

INFLUENCE OF HYDROGENOUS SUBSTANCES ON THE RADIOACTIVITY PRODUCED BY NEUTRONS

In performing some experiments on the neutron-induced radioactivity of silver we noticed the following anomaly in the intensity of the activation: A layer of paraffin a few centimeters thick inserted between the neutron source and the silver increases the activity rather than diminishes it. After this experiment we ascertained that the presence of large paraffin blocks surrounding the neutron source and the target increases the activation intensity by a factor which, depending on the geometry used, ranges from a few tens to a few hundreds.

After these observations we tried to ascertain—for the time being, only summarily—the circumstances under which this phenomenon occurs. The facts which have been ascertained up to now are the following:

a. A radium source, without beryllium, does not produce the effect —a circumstance which makes us attribute the effect to neutrons and not to gamma rays.

b. An effect of approximately the same intensity as with paraffin is obtained with water. We deem it very probable that the effect depends on the presence of hydrogen because substances containing oxygen but not hydrogen ($NaNO_3$) do not produce an increase of activity, at least of the same order of magnitude.

c. The effect observed in silver is not shown by all elements which can be activated by neutrons. We have observed, up to now, that silicon, zinc, and phosphorus do not show an appreciable increase in activity, whereas copper, silver, and iodine give rise to activities much larger than one would obtain without the presence of water.

Professor at Rome

From these few cases it seems the rule that the only elements sensitive to hydrogenous substances are those which under bombardment give rise to activities due to isotopes of the starting element.

The case of aluminum is noteworthy. In water it acquires an activity that shows a period slightly shorter than 3 minutes, corresponding to that of Al^{28}, which can be extracted from irradiated silicon. This activity under normal conditions is so weak that it almost disappears in comparison with other activities generated in the same element. Similarly, zinc and copper, which give rise to the same active products,* (radioactive isotopes of copper), under normal conditions acquire activities of the same order of magnitude; but, in water, copper shows a much larger effect than zinc.

A possible explanation of these facts seems to be that neutrons rapidly lose their energy by repeated collisions with hydrogen nuclei. It is plausible that the neutron-proton collision cross section increases for decreasing energy, and one may expect that after some collisions the neutrons move in a manner similar to that of the molecules of a diffusing gas, eventually reaching the energy corresponding to thermal agitation. One would form, in this way, something similar to a solution of neutrons in water or paraffin surrounding the neutron source. The concentration of this solution at each point depends on the intensity of the source, on the geometrical conditions of the diffusion process, and on possible neutron-capture processes due to hydrogen or to other nuclei present.

It is not ruled out that this point of view may be important in explaining the effects observed by Lea.†

The investigation of all these phenomena is being continued.

E. Fermi, E. Amaldi, B. Pontecorvo, F. Rasetti, and E. Segrè
Istituto Fisico della R. Università.
Rome 22 October 1934–XII.[34]

* T. Bjerge and C. H. Westcott, *Nature* 134 (1934): 286.
† D. E. Lea, *Nature* 133 (1934): 24.

The discovery of the hydrogen effect opened a host of problems and made us reorient our entire research program. First we measured, for many substances, what we called the coefficient of aquaticity: how much the immersion in water would increase activity under specified conditions. This measure confirmed that the (n, γ) reactions were the only ones sensitive to hydrogenous substances, and by early November we were convinced that a slowing down of the neutrons was the correct explanation of the phenomena. Our attention then turned more to the study of the slow neutrons than to that of the substances produced, and we tried to see whether slowing down in hot or in cold water would change the properties of the neutrons. We believed that neutrons were effectively thermalized, but at first we could not show any temperature effect. Nevertheless, we kept trying. Ultimately, P. B. Moon and J. R. Tillman in England demonstrated a thermal effect, and we hastened to repeat their experiments. We soon found that some substances—cadmium, for instance—strongly absorb slow neutrons, and we measured crude cross sections for this effect. We detected the gamma rays emitted when the neutrons are captured and we started crude measurements of the density of slow neutrons in a hydrogenous medium as a function of the distance from the source. Finally, we tried to slow down neutrons by collisions with substances other than hydrogen and found some effects of inelastic collisions that explained our original observations on the effect of surrounding source and detector with a lead housing. All this work had produced significant results by December 1934, that is, about six weeks after the discovery of the slow neutrons.

I have dealt at some length with these investigations in an effort to recreate the spirit of that heroic pioneering period.

Immediately after the discovery of slow neutrons, Corbino suggested that they might have important practical applications and that it would be advisable to take out a patent on this work. This was done and resulted in the Italian patent no. 324 458, of October 26, 1935. This patent covered a process for the production of radioactive substances, in particular by neutron bombardment, and covered also the enhancement of the effect by slowing down the neutrons by multiple collisions. Because slow neutrons are the agent operative in nuclear reactors, the patent has become basic to the production of nuclear power. It is also basic to military applications, which use both fast and slow neutrons. The original Italian patent was in the course of time subdivided for technical legal reasons and extended to other countries. The inventors are Fermi, Amaldi, Pontecorvo, Rasetti, and Segrè. We agreed, however, to share equally any possible benefit with D'Agostino and Trabacchi.

The steadily deteriorating political situation in Europe suggested that we try to transfer our interest to an American company. We accordingly got in touch with our friend G. M. Giannini (not to be confused with Giannini of the Bank of America), who had studied physics in Rome and had emigrated to the United States. With him we made an agreement whereby he would hold the patent and share equally in it with the inventors. Legal expenses for the extension of the patent to other countries were paid by the Philips Company of

Professor at Rome

Eindhoven, The Netherlands, which in consideration of this action also acquired a share in the patent. Attempts to interest American companies (such as General Electric) in the possibilities of practical applications for nuclear physics failed completely despite Fermi's presentations to their experts—a remarkable contrast in farsightedness between the relatively small Philips Company and the American industrial giants. Fermi's presentation was ineffective not because of a difference in business outlook between him and the American businessmen—as I thought at the time—but because the industrial tycoons simply failed to grasp the practical potentialities of the new field of nuclear physics that was then opening up. Lewis L. Strauss, a well-known American investment banker who later became the chairman of the Atomic Energy Commission, as late as 1938 was unable to interest his personal and business friends in nuclear applications. His efforts came to naught, he said, because "the top management of the company in those days felt that nuclear energy was 'for the science-fiction fans.' "[35] When fission was discovered and atomic energy developed, the patent became basic to this work and thus acquired great value, especially in the eyes of those who knew of the secret developments in the United States.

Wartime, for obvious reasons, was an inappropriate period for business transactions with the government. It was practically impossible and also undesirable to raise any questions relative to the patent at a time when its subject was involved in a project vital to national security. At the end of the war there was a long period of uncertainty while Congress debated the Atomic Energy Act, which had to deal specifically, among other things, with just compensation for patent rights belonging to private individuals and preempted by the government. Because the invention was due largely to Fermi's efforts the business decisions were left mainly to him. After the end of the war, during the period of uncertainty about the government's role in atomic energy matters, the United States government bargained very hard with Fermi, who probably would have renounced his rights had he not felt an obligation to the other inventors. The government lawyers used every possible reason to reduce and delay the compensation. Among other things they advanced the argument that Fermi was now a member of the General Advisory Committee of the Atomic Energy Commission. Although the members of this high advisory board (which gave technical guidance to the AEC) were not paid, the lawyers contended that there was a conflict of interest and that for this reason neither Fermi nor the other inventors could claim just compensation for their patent rights. This seriously annoyed Fermi. He had been most generous in assigning other patent rights to the government for work done at Columbia and for which he could have claimed, and most probably have obtained, substantial compensation. The patent under question referred to work done in Italy, in 1934, by a group of which he was only one and which had no connection with the AEC or any of its predecessors. The legal pettiness disgusted him.

While the dealings with the government lawyers, who seemed principally interested in delaying any conclusion, slowly proceeded, the situation was further complicated by the mysterious disappearance of one of the inventors, Pontecorvo, who as it turned out later, had fled to the Soviet Union in September of 1950.

Giannini, who as assignee of the patent had asked the United States Court of Claims for compensation of $10 million (although this sum seemed unrealistically high to the inventors), lost heart and withdrew the suit. He then announced in the press that he had divested himself of all patent rights, (although he later collected his share). Finally a settlement was reached, in the summer of 1953, whereby the government paid "just compensation" of about $400,000. The share of each inventor after the legal expenses were paid—and Fermi's share as well—was approximately $24,000.

The discovery of slow neutrons in 1934 was equivalent to multiplying the strength of our sources by a substantial factor; it allowed us to see or reexamine phenomena that had been too weak for study only a few months before. One of the first of these to be investigated again was the (n, α) reaction on boron. More systematic work on uranium followed, but without much progress.

Another unsuccessful attempt was made in January and February of 1935 to explain the great number of new activities induced in thorium and uranium, which had been isolated by us as well as by other groups. We tried to interpret all the observed activities as due to transuranic elements and their possible decay products, and we thought that, besides beta emitters, there must be alpha emitters. Thus we began to irradiate foils of uranium or thorium (in the form of oxides), and we put them, immediately after irradiation, in front of a thin-window ionization chamber, connected to a linear amplifier. As we failed to observe any new alpha activity, we thought that the activities might have too short a half-life for our experimental arrangement—perhaps a fraction of a second. Therefore we set the uranium in front of the ionization chamber and irradiated it with a neutron source that was surrounded by paraffin. We thought that if the alpha emitters had a short half-life, they ought to emit (according to the Geiger-Nuttall law) alpha particles of a range considerably greater than that of the particles emitted by uranium or thorium. For this reason the experiments were always carried out with uranium or thorium covered by an aluminum foil equivalent to 5 or 6 cm of air and gave a negative result. It was this aluminum layer that prevented us from seeing the big ionization pulses characteristic of fission, but it is impossible to say whether we would have correctly interpreted the phenomenon had we observed it.

The main results of all our work were summarized in a second paper for the Royal Society, written in February 1935,[36] which contained the seeds of most of the important ideas and facts of neutron physics. Fermi had by this time made substantial progress in the theory of the slowing-down process of neutrons, but nuclear physics had a serious difficulty in the apparent lack of correlation between the scattering and the capture cross sections in cases where the capture cross section was very large, as with cadmium. Fermi made several attempts to explain this phenomenon, but its mechanism was made clear only through Bohr's work on the compound nucleus. On the other hand, the concept of scattering length and the possibility of resonances were very much present in Fermi's mind. The concept of scattering length, with its typical diagram, had been developed by Fermi in 1934 to explain the pressure shift of spectral lines near the limit of series that Amaldi and I had discovered.[37] Much to our surprise, Fermi, with his uncanny ability for discerning analogies in in phenomena that seemed to be completely unrelated, transferred the theory of the shift of the spectral lines to neutron physics. He also made a number of calculations on the behavior of neutrons in hydrogenous media by what we would today call the Monte Carlo method—that is he followed in detail the fate of a neutron in its successive collisions, determining by chance (Monte Carlo) the parameters of each collision. Having followed many neutrons he could make a statistical study of the results. He did not speak about his method then, and I learned that he had used it in 1935—only many years later, in Los Alamos, from Fermi himself. In the spring of

Professor at Rome

1935 we devised a mechanical experiment by which we could compare neutron velocities with the peripheral velocity of a wheel. We began to use our rapidly rotating wheel only a few days before Fermi departed from the Ann Arbor summer school, and the results were not statistically established until his luggage was packed and he was about to take the boat from Naples. This is the only "rush case" I remember of all our work in Rome, which, for all its dramatic moments, was very intensive but never hurried.

When we left for our vacations in the summer of 1935, our mood was not happy as in previous years. Political developments, especially the preparations for the Ethiopian war and the deterioration of the situation in Europe, had seriously affected our work. I was very conscious of the change and talked about it with Fermi, who said I would find the answer in the library of the institute on the big table, in the center of the reading room. I went to the library and found a world atlas lying on the table. The atlas had been consulted so frequently that when opened at random it invariably showed the map of Ethiopia. By the fall of 1935 the crisis was at its height. Fermi had returned from Michigan, but the group had dispersed. Rasetti, who had gone for the summer to Columbia University, decided to stay longer. I had been appointed professor of physics and director of the laboratory at Palermo and had moved there. Pontecorvo soon left Italy for France and D'Agostino took a position with the Consiglio Nazionale delle Ricerche. The atmosphere in Italy was gloomy as the country prepared for war and the League of Nations imposed sanctions against her.

Amaldi, who worked in Rome with Fermi during 1936, described the events of that year:

After the summer vacation of 1935 Fermi and I found ourselves alone in Rome . . . and we turned our attention to some results of Bjerge and Westcott and of Moon and Tillman, who had observed that the absorption of slow neutrons by various elements differed slightly, depending on the element used as detector. This fact was not explained by the current theory of the absorption of neutrons by nuclei. The theory predicted a capture cross section inversely proportional to the velocity of the neutrons for all nuclei. This energy dependence was supposed to be valid for such large energy intervals as to cover the entire energy range of slow neutrons.

We went to work with even greater energy than in the past, as if by our own more intensive efforts we wanted to compensate for the loss of manpower in our group.

We had prepared a systematic plan of attack that we jokingly summarized by saying that we would measure the absorption coefficient of all 92 elements, combined in all possible ways, with the 92 elements used as detectors. In jest we added that, after combining all the elements two by two, we would also combine them three by three. By this we meant that we would also study the absorption properties of the neutron radiation filtered in several ways.

Actually, after having measured the absorption coefficient of eleven different elements in all possible combinations with seven detectors, we were convinced that the observations of the two English research groups were correct and, in general, that the rule that the absorption coefficient of a certain element was greater when the element itself was used as detector was valid. We began to study the particular cases of silver, rhodium, and cadmium in great detail. . . .

I think it is interesting to note that on this occasion I, as on many others, tended to form a simple picture of the phenomenon; I tried to interpret the different groups of neutrons as different bands of energy. Fermi, however, did not want to accept this description. He too was convinced that this was obviously the simplest hypothesis, but he maintained that it was not strictly necessary, at least for the moment, and was therefore harmful if introduced into our mental picture. He insisted that one must proceed by reasoning with the observed experimental facts. The correct interpretation of the nature of the neutron groups would emerge as a necessary consequence of the data. He was afraid that a preconceived interpretation, however plausible, would sidetrack us from an objective appraisal of the phenomenon that confronted us.[38]

This reaction is peculiar, but the same attitude appears in a speech Fermi gave in 1951:

> One can go back to the books on method (I doubt whether many physicists actually do this) where it will be learned that one must take experimental data, collect experimental data, organize experimental data, begin to make a working hypothesis, try to correlate and so on, until eventually a pattern springs to life and one has only to pick out the results. Perhaps the traditional scientific method of the textbooks may be the best guide, in the lack of anything better.[39]

I suspect that Fermi made such utterances with tongue in cheek. I believe that he always had a working hypothesis but wished to test it thoroughly before mentioning it. Perhaps he even distrusted and tried to fight his own scientific conservatism; he did not want to miss great discoveries by not being bold enough. His reaction to Heisenberg's early work on quantum mechanics, the events connected with the discovery of fission, and the success of the beta-ray theory must

Professor at Rome

have taught him the pitfalls of excessive conservatism. On the other hand, he did not want to be rash.

At any rate, Fermi and Amaldi began a systematic study of the absorption properties of the various neutron groups and labeled them with letters, both for brevity and to avoid committal to an interpretation. Group C, strongly absorbed by cadmium, proved to be of truly thermal energy, a result they obtained by showing that all other groups could be transformed into group C by collision. The explanation of neutron groups as energy differences of resonant absorption lines was established by these experiments and explained by the work of Bohr on the compound nucleus. Bohr's ideas as well as Breit's and Wigner's work provided explanations for the narrowness of the resonances and the density of levels for intermediate and heavy nuclei. Fermi and Amaldi measured the width and location of resonances as well as they could with the methods available to them. Some of these methods were extremely ingenious and profoundly influenced later developments in neutronology; moreover, the techniques were considerably improved over those that had been used the previous year. Ionization chambers connected to electrometers replaced the Geiger-Müller counters and could measure radioactivities to an accuracy of a few percent. Calibrated sources, nomograms for reducing measurements, and standardized procedures produced a precise and rapid technique.

Many important measurements were summarized in an experimental paper, "On the Absorption and the Diffusion of Slow Neutrons," by Amaldi and Fermi (*FP* 118), which was translated into English by Amaldi and published in the *Physical Review*. A companion theoretical paper by Fermi, "In the Motion of Neutrons in Hydrogenous Substances" (*FP* 119), gave the theoretical bases of the experiments discussed in Amaldi's paper and in many ways established the foundations for neutronology. Fermi published his paper only in Italian and said that people who wanted to work on neutrons would have to read it anyway, so why bother to translate it. In fact it was translated into English several years later, when the motion of slow neutrons in moderators became a hot technical subject.

Amaldi, who was very close to Fermi throughout this period, has summarized the events of the academic year 1935–36:

> The academic year, which had slipped by in an atmosphere of frenzied work and isolation, was drawing to a close. Rasetti wrote to us every now

Professor at Rome

and then about what was happening at Columbia University, reprints sent by Halban and Preiswerk kept us informed on the work in Paris, and correspondence with Placzek kept us in contact with Copenhagen. Through this latter correspondence we learned of Bohr's work, as well as that of Frisch and Placzek, on the $1/v$ absorption law in boron. Through this correspondence the joke spread from Rome that, as the captain's age can be determined from the length of the ship's mast, the energy of a neutron group can be determined by the distance it travels as it slows down. The term "age," used later by Fermi to represent the magnitude $<r^2>/6$, might date to this period. At first the expression "the age of the captain" was used to refer to experiments on the transformation of one group into another group of lower energy. . . .

We worked with incredible stubbornness. We would begin at eight in the morning and take measurements, almost without a break, until six or seven in the evening, and often later. The measurements were taken with a chronometric schedule, as we had studied the minimum time necessary for all the operations. They were repeated every three or four minutes, according to need, and for hours and hours for as many successive days as were necessary to reach a conclusion on a particular point. Having solved one problem, we immediately attacked another, without a break. 'Physics as soma' was our description of the work we performed while the general situation in Italy grew more and more bleak, first as a result of the Ethiopian campaign and then as Italy took part in the Spanish Civil War.[40]

In Aldous Huxley's novel *Brave New World* soma pills based on a sexual hormone were taken by men of the year 2000 to fight despondency.

The idea of using "physics as soma" apparently is quite familiar to physicists; I saw it applied extensively in the tragic years of the Nazi's domination of Germany, and Fermi recommended it to me at a time when I was discouraged by failure in a competition. When the Russian physicist Kapitza was detained by the Soviet authorities in Russia where he was vacationing and prevented from returning to his post at Cambridge, England, Lord Rutherford wrote him a remarkable letter, in which "soma" is prescribed.[41] Recourse to work as "soma" is not new and certainly is not peculiar to physics; no doubt it is effective.

There was a marked contrast between the success of the important work accomplished in 1936 and the simultaneous deterioration of the political situation in Europe, foreboding a bleak future for all of us. New travels to the United States were indicated, and Fermi went to Columbia University for the summer session. He gave a course on thermodynamics which formed the basis for a small book that is still

Professor at Rome

popular.⁴² Amaldi and I too visited the United States, and in the fall we all returned to Italy.

For us the general political troubles were further aggravated by a totally unexpected blow. On January 23, 1937, Professor Corbino died of pneumonia after a short illness. He was only sixty-one years old, and we had expected that we would have his assistance and guidance for many more years in the difficult times we foresaw. It would have been natural to appoint Fermi as Corbino's successor; however, by maneuvering in the science faculty, Lo Surdo became the new director of the institute in Rome. This was a sign that Fermi's fortunes were declining in Italy; furthermore, it did not augur well for the continuation of the work in Rome. Actually, however, in the next two years Lo Surdo did not exert much influence.

Fermi wrote a brief eulogy of Corbino in which characteristically he emphasized Corbino's scientific work, but in which he also showed his affection and appreciation for Corbino's warm personality.⁴³

Some time earlier Amaldi had won a competition for a chair of experimental physics in Sardinia but did not go there; now, upon Corbino's death, he was given the chair in Rome. Amaldi was only twenty-nine years old, and the appointment was unusual for a person of his age; however, it was an excellent choice and worked to the great advantage of Italian physics. It was also in 1937 that the institute was moved from its old building in Via Panisperna to the new University City. The new quarters were much larger and on the whole better, but the move itself caused considerable disruption, as was to be expected for the transfer of such a complex organization as a scientific institute.

Before moving, Amaldi, Fermi, and Rasetti started to build a small Cockcroft and Walton accelerator that would produce 200 kev deuterons, because the radon-plus-beryllium sources, despite their advantages of constancy and small geometry, could not compete with the burgeoning accelerators.

For all its great success, the neutron work started in 1934 had an adverse side effect on the very spirit of the Rome institute: the work became so absorbing and time-consuming that it drastically changed the group's working habits. Fermi could not pay as much attention as previously to students and visitors and to developments in physics outside the field of his immediate interest. There was no time for

Professor at Rome

leisurely study of the issues of *Zeitschrift für Physik* or for discussions with young physicists from abroad. Gone were the seminars and private lectures from which in previous years we had learned so much. The study of physics became narrower and more utilitarian than it had been because to keep our leading position in neutron physics we had to accomplish a tremendous job as fast as possible. (Little did we suspect that this way of working would extend through a great part of our scientific lives.) For Fermi, especially, the load steadily increased as time went on; the pressure of work—or perhaps the mere passage of time and his growing maturity (after all he was only in his thirties)—seemed to affect him in another way: he became more reserved in his personal relations, more reticent. The tendency to be discreet and to keep his own counsel increased with time, and later was enhanced by the demands of war work in the United States. Despite all his cordiality and apparent ease of communication, I believe that Fermi was never prone to confide his innermost thoughts, either scientific or otherwise, except after very mature reflection.

Another change in the world of physics was the eclipse of Germany; this was very apparent to us because we had been in Germany at one time or another and had many friends among the German scientists. Because the Nazi blight was destroying physics in Germany and we were fearful the contagion might spread to the rest of Europe, we turned more and more to England and the United States. Our friendliness for England was due in part to the important role Rutherford and his school had played in the development of nuclear physics, and America looked like the land of the future, separated by an ocean from the misfortunes, follies, and crimes of Europe. Thus we started to learn or improve our English, and Fermi decided to publish the important papers on neutrons in English rather than German. This change was instigated and wholeheartedly approved by his collaborators who had a profound resentment against the Nazis and their crimes. We had a pretty clear idea of what was occurring in Germany, but at that time the crimes were minor compared with what was to come. The situation, of course, did not affect our personal relations with our German friends, the great majority of whom were victims of a tyranny they hated as much as we did. We maintained cordial relations with Hahn, Meitner, and others, and until about 1936 exerted our very modest power to help our friends.

Professor at Rome

After that time it became practically impossible to do anything from Italy.

To understand Fermi's political attitude during his Italian period we must remember that he came from a middle-class family of civil servants and that he was educated in Italian public schools. The liceo gave a humanistic varnish of culture called by a distinguished contemporary Italian historian "the illiteracy of the literates." It was essentially a superficial literary and historical glorification of Italy in which the Roman Empire was presented as direct ancestor of modern Italy and the recent Risorgimento was transformed into a nationalistic myth. Fermi was too critical even as a boy to believe all that he was taught, but it was difficult for a young man isolated in provincial Italy to broaden his outlook sufficiently to grasp world events and their significance. Corbino was almost the only important man of affairs he knew in his formative years. Later he rapidly absorbed the culture of the countries he visited.

When fascism came to power in Italy in 1922, it had the support of a very large portion of the middle class, which interpreted its advent as a return to order after a period of turmoil. Furthermore, fascism appealed to the patriotic and nationalistic feelings that the secondary schools had implanted in most of the Italian youth. The rest of the population was divided; some, like Professor Volterra, saw immediately the long-range implications of Mussolini's seizure of power; others, like Corbino, were ambivalent; he never joined the Fascist party, but he served in Mussolini's early coalition cabinets.

The first severe shock to fascism's early sympathizers was the infamous murder of Matteotti (in 1924) and its consequences. This was the turning point when many people changed allegiance. Fermi never participated actively in politics, although he took an intellectual interest in it. He was extremely busy with his own work and had little time to spend in other activities. Fermi's characteristics—sincerity, integrity, and a tendency to understatement—were incompatible with the bombastic rhetoric of fascism. Aware of many evils of the Fascist government, he did not actively support it; on the other hand, he never participated in militant antifascism. As long as he could preserve his personal integrity and work unhindered in physics, he tried to ignore the vagaries of the government under

Professor at Rome

which he found himself. He did not like to engage in political arguments nor waste his energies on things that were beyond his control.

With his appointment to the Italian Academy in 1929, Fermi found himself automatically among Fascist bigwigs, but he shunned public functions as much as possible and kept away from the unsavory crowd which surrounded the government officials. He would not use his official position, much less his scientific preeminence, to obtain favors for anybody, and least of all for himself. An anecdote is illustrative of his attitude. In 1930, when the crown prince of Italy married, high state officials, including Fermi, were invited to the wedding and given a special card of admission for it, but Fermi decided to work in the laboratory instead. To get to the laboratory he had to cross a street on the procession route that had been closed to traffic and was guarded by lines of soldiers. Fermi, driving his shabby little car in his usual clothes instead of the brilliant uniform of the academy, nevertheless had the invitation card in his pocket, and when stopped by the soldiers, he showed it to an officer. "I am the chauffeur of His Excellency Fermi," he said. "I have to fetch him for the wedding. Could you please let me cross the soldier's lines?" Whereupon he was led through the lines and spent the rest of the day at work in his laboratory.

Hitler's rise to power in January 1933 was a major turning point in European affairs. Everybody who had knowledge of or contact with Germany could see that dire events were in the offing; but some hoped the situation would correct itself, while others tried not to see the unfolding truth because it was too horrible to contemplate, and many lost their lives because of blind optimism. Mussolini, originally unfriendly to Hitler and alarmed by the turn of events in Germany, correctly considered Hitler a dangerous man and for a certain period hoped that it would be possible to check him. At the time of the abortive putsch in Austria and the murder of Chancellor Dollfuss in the summer of 1934, he went so far as to send Italian troops to the border in a demonstrative action intended to intimidate and stop Hitler. Later, ensnared by his own rhetoric, unprepared and misinformed, he launched the Ethiopian adventure. He thought he could gain the acquiescence or the neutrality of France and England by taking their side against Hitler. This cynical gambit failed, and the final outcome of the maneuvering was that Mussolini and Italy slipped more and more into the German orbit. Fascist intervention in the Spanish Civil War was another step in a catastrophic course.

Professor at Rome

Mussolini, who initially had delusions of being the more important of the two dictators, was continuously forced into the position of junior partner, and finally into that of Hitler's vassal.[44]

The first manifest giant step in this abysmal descent was the formation of the Rome-Berlin "Axis" in November of 1936. Then Italy's acquiescence in the Anschluss—that is, Germany's subjugation of Austria—in March 1938 demonstrated that the process of degradation was almost complete. Mussolini and the Italian government had been willing and able to stop Hitler's attempt to annex Austria four years earlier, but now they were powerless to do anything, and to conceal their impotence they gave their approval to the deed.

After the Anschluss Erwin Schrödinger had to flee Austria, where he was a professor. He escaped on foot with nothing but a rucksack and came to Rome. Immediately, he went to Fermi and asked to be accompanied to the Vatican to seek protection. This event and other ominous daily portents deeply disturbed those physicists who were still in Rome.

Hitler himself came with great pomp to visit Rome as an official guest in May of 1938, and the visit indicated what could be expected. A bitter epigram by the poet Trilussa, circulated by word of mouth, reflected in its lines the feeling of many, in particular of Fermi and his entourage:

> Roma di travertino
> Rifatta di cartone
> Saluta l'imbianchino
> Suo prossimo padrone
>
> Rome of travertine splendor
> Patched with card and plaster
> Welcomes the little house painter
> As her next lord and master[45]

The allusions are to the patch job ordered by Mussolini to conceal slums on the route to be followed by Hitler and to the early trade of his guest.

Mussolini's theatrical rhetoric on the subject of "Impero"—the idea of presenting the precarious occupation of Ethiopia as a restitution of Roman glories—was repugnant to persons of cool and honest judgment and especially to scientists. The fascist government started also to promulgate strange laws penalizing bachelors and prescribing ways of speaking and of dressing. Italy's alliance with a Germany bent on a criminal course was even more disgusting. Fermi's feelings were expressed in the bitter remark that Italy would be saved if Mussolini became overtly insane and walked on all fours in the Piazza Venezia—under the balcony from which he usually spoke.

The final demonstration of Italy's ideological enslavement was the promulgation of the laws against the Jews. These laws personally affected Fermi because his wife was Jewish. Italy's anti-Semitic campaign was initiated by the *Manifesto della Razza* of July 14, 1938, which was concocted by members of the government and possibly by Mussolini himself. It purported to give scientific reasons why Italian Jews should be considered aliens and persecuted. The government tried to make the infamous doctrine respectable by having "scientists" and intellectuals sign or indorse it, but only government sycophants put their names to it and in later years some of these disclaimed ever having signed it. To the credit of the Italian universities, Mussolini found only five professors who were willing to put their names to the document. There were numerous acts of protest and expressions of sympathy for the Jews, and many civil servants who found conflicts between their official duties and their consciences let their consciences rule their actions. The theoretical expression of the racial doctrine was soon followed by specific laws copied from the Nazi Nuremberg edicts. The 1938 laws did not affect Fermi or his two children, who had been baptized and were considered Aryans, but his wife's family was severely affected and it was not clear what the laws might bring to his wife. The danger was not immediate, but the racial laws were so revolting to Fermi's sense of justice and indicated such a degeneration of the fabric of Italian life that he decided to leave.[46]

In previous years he had received several offers from foreign universities—he had, for instance, been offered the professorship at Zurich vacated by Schrödinger in 1928, a position at the Institute for Advanced Study at Princeton, and positions at several other universities in America. On these occasions he had seriously considered leaving. Fermi himself was strongly attracted to America, but his pupils and friends urged him not to go, and his wife was reluctant to leave Rome. Carefully weighing his responsibilities to his family, friends, and pupils, he had decided to stay. Now, however, the circumstances were different and precluded further hesitation. He wrote to four American universities that had approached him

Professor at Rome

previously and told them that the reasons that had prevented his accepting the offers were no longer valid. He received prompt answers from all of them and decided to accept a professorship at Columbia University.

In the fall of 1938, at a physics meeting in Copenhagen, Fermi was confidentially informed by Bohr that his name was high on the list of candidates for a Nobel prize. Such a disclosure was most unusual but was prompted by an effort to ascertain that the award would not embarrass him. There had been cases in which dictatorships had forced a recipient to renounce the award or had otherwise harassed him. The Swedish Academy was thus taking precautions. The nomination to the award was hardly surprising, but it called for some adjustments in Fermi's emigration plans. He wanted to proceed directly from Stockholm to New York without returning to Italy, to avoid possible reprisals, such as the refusal of passports for the family, and the laws that would have frozen the award money.

He let it be known that he would go for a visit to Columbia University, and because he had previously traveled so often to American universities, the authorities might regard his trip as routine, especially if they did not wish to be too zealous. He secured from Columbia a "public" invitation for a seven months stay, a period that would necessitate an immigration visa (rather than a tourist visa), which he obtained under section 4D of the Immigration Act of 1924, allowing professors to obtain an immigration visa outside the quota system.

Once Fermi made a decision, he never looked back. He told a few close friends, including Amaldi and Rasetti (I was already in Berkeley), that he was about to emigrate permanently, but he would not discuss the resolve. Amaldi's wife, Ginestra, was particularly distressed; she did not think the move was justified and suggested that it resembled desertion in a difficult situation. Edoardo Amaldi, though deeply regretting the dissolution of the Rome group, realized that the move was unavoidable.

On November 10, 1938, Fermi was notified of the award of the Nobel Prize in physics. The public announcement followed shortly, but the Fascist press, usually anything but indifferent to national glories, published the news in a remarkably inconspicuous manner. At a time when the press was entirely controlled by the government, the Fascists obviously had decided to underplay the event. This was

Professor at Rome

not surprising: Fermi had been personally attacked by extremist Fascist newspapers for "having transformed the physics institute into a synagogue," and the same issue of the newspaper that reported Fermi's Nobel award carried prominently the news of a second batch of racial laws.

On December 6, 1938, the Fermi family—Enrico, Laura, the two children, Nella and Giulio—and a maid boarded the train at Rome. Amaldi and Rasetti went to the station to say goodbye. Everyone understood that the departure signified the end of a memorable period in their lives.

The Rome group was in effect dissolved by uncontrollable forces. I had gone to the United States in July 1938 for a visit and, when I was dismissed from my Palermo post, settled in Berkeley. Fermi left Italy in December of 1938. In the summer of 1939 Rasetti emigrated to Canada, and Amaldi came to the United States to look for a job, although his wife was reluctant to leave Italy. He applied for a passport for his family, but without waiting for it he left Italy alone to explore American possibilities. Before he could find a suitable job, however, Germany invaded Poland, passport for the family was refused, and he returned to Italy in October 1939, after having spent some time with me in Berkeley and with Fermi in Ann Arbor and Leonia, New Jersey.

When Amaldi returned to his post in Rome, he had to face heavy and unexpected responsibilities. The history of Italian physics during the war period is beyond the scope of this book; a large part of the task of maintaining the international position that Italian physics had acquired during the decade of Fermi's preeminence devolved upon Amaldi, who succeeded admirably. That this was at all possible shows the depth of Fermi's influence on Italian physics.

At Stockholm, Fermi went through the colorful traditional ceremonies of the Nobel award and received the prize on December 10, 1938, from King Gustav V. The official speech was given by Professor H. Pleijel of the Swedish Academy, who concluded with the following words (pronounced in Italian):

> The Royal Academy of Science of Sweden has awarded you the Nobel Prize in physics for 1938 for your discovery of new radioactive substances belonging to the entire range of elements and for the discovery you made in the course of this work of the selective power of slow neutrons.

Professor at Rome

We offer you our congratulations and express the highest admiration for your brilliant researches, which throw new light on the constitution of atomic nuclei and open new horizons for the further development of atomic investigation. I beg you now to receive the Nobel Prize from the hands of his Majesty the King.

Fermi, following tradition, gave a lecture on the work for which he had received the prize. (The Nobel lecture—a semipopular description of the entire Roman neutron work—is reproduced in Appendix 2.) His speech, given before the discovery of fission, contains the names "ausonium" and "hesperium" for elements 93 and 94. Fission was discovered between December 10, 1938, and the printing of the speech, and is mentioned in a footnote. As is well known, the discovery of fission required a reappraisal of the work on the radioactivities induced in uranium. It has occasionally been said that Fermi's was the only Nobel Prize awarded for an unsubstantiated discovery—on the assumption that discovery of the transuranic elements was the reason for the prize. It is clear from the presentation statement, however, that this was not the case.

While the Nobel festivities proceeded in Sweden, a great discovery was being made in Germany. Otto Hahn and Fritz Strassmann were establishing beyond doubt the presence of radioactive barium among the products of the neutron irradiation of uranium. Their paper was sent to *Naturwissenschaften* on December 22, 1938.[47] This was to be the last great achievement of German science for many years, the last ray of sunset before the setting in of night.

On being informed of Hahn's discovery, Frisch and Meitner almost immediately surmised that the presence of barium among the isotopes produced by the neutron bombardment of uranium indicated fission, the splitting of the uranium nucleus into two large fragments. Within a few days this interpretation was confirmed by the observation of the large ionization pulses produced by the fragments, and the publication of these results followed on January 15, 1939.[48]

The news of these discoveries was propagated by word, letter, and telegram while it was in the making, and before it ever appeared in print. The news produced a major sensation among the initiates, but Fermi heard nothing of it during his last days in Europe. After the Swedish ceremonies, the Fermis spent a few days in Copenhagen, where they were cordially received by Bohr and his family, and on December 24, 1938, they boarded the *Franconia* at Southampton.

Professor at Rome

The Fermis landed in New York on January 2, 1939, and Fermi turned to his wife and said: "We have founded the American branch of the Fermi family."[49] (I can imagine his smile.) Professor Pegram, chairman of the Physics Department at Columbia, and G. M. Giannini greeted them on the pier.

4
EMIGRATION AND THE WAR YEARS

In 1939 Fermi was at the midpoint of his career and on the verge of becoming involved in events whose historical importance would transcend anything he could have anticipated a few years earlier; it is therefore important to try to describe his outlook and the way he typically reacted to events at this time. What follows is a subjective analysis that gives only my impressions but which may help in interpreting Fermi's behavior.

The strict discipline of the severe, industrious, middle-class family in which Fermi had been brought up left a deep imprint. His life—except for his all-important science—was patterned on that of an efficient, loyal civil servant. In his early youth he had not been exposed to a broad view of current events, and his first contacts with a wider world and with social or political issues came later, mainly through Corbino. In his last years in Italy, Fermi served in a few high advisory posts in the Ministry of Education and as consultant for industrial firms, but although he always discharged such duties conscientiously, he considered them minor occupations accepted either because he felt it was his duty to do so or because they supplemented his income. When Corbino's death produced a vacancy on the board of directors of a large corporation, the seat was offered to Fermi. This would have been a rare opportunity of gaining an important position in the business world, but he politely declined the offer: it simply held no interest for him.

The War Years

Fermi abhorred battles, and especially those whose outcome was uncertain. He often said that one should avoid lost causes; Don Quixote was not his hero. He was extremely fair, judicial, and impartial in all his actions, striving always to see other persons' point of view and to avoid favoritism or even the slightest semblance of it. But for all his love of justice he did not seem to be emotionally involved in abstract causes. He carefully avoided entanglements in issues that were not concrete and definable and which did not offer a good probability of a favorable outcome. When he was confronted with superior forces that were clearly beyond his control, he would retreat, either by withdrawing from the problem or, in personal matters, by avoiding the individuals involved. Possessing great discipline and natural reserve, he seldom commented on persons or actions he disliked. (Thus, even knowing him well and seeing him frequently, I never learned some of his opinions on important political events and persons.)

Fermi relied upon simple principles but used his extraordinary analytical powers to build on them. He was fully aware of his extraordinary mind, and I have the impression that he sometimes mused on his role in science and on his place in the history of science. In some ways he felt himself to be above the transient noise and excitement of current events. He was mindful of eternity.

In scientific matters he was conservative. He hated to promise more than he could deliver or to conclude from an experiment or calculation more than the results actually indicated. He was always cautious in assigning probable errors to his results and he often, half-jokingly, referred to the *culto dell'ipotesi cauta*, the "cult of the cautious hypothesis" (a phrase that had been coined to bestow faint praise on an Italian physicist who had achieved very little). All this goes to show that Fermi was prone to understatement, a fact that is also evident in his scientific papers. Few things caused him more anguish than the unexplained behavior of the transuranic elements before the discovery of fission revealed the true nature of the phenomena. The following anecdote shows how deeply he felt about his failure to solve this problem. One day after the war Fermi and some of his colleagues were studying the architect's sketches for the future Institute for Nuclear Science at the University of Chicago. The drawings showed a vaguely outlined human figure in bas-relief over the entrance door. When a group began speculating as to the significance of the human figure, Fermi immediately interjected that it was probably "a scientist not discovering fission."[1]

The War Years

Because Fermi loathed being in error, and error is occasionally unavoidable, he wanted to be in error only for having claimed too little. In the privacy of his thoughts he would make daring hypotheses, but he would never publish them until they had been firmly substantiated. Similarly, he would never give anybody a promise or even a hope of anything unless he was sure he could deliver it.

His interests outside physics were rather limited. He did not cultivate music (except for playing the piano with two fingers), and he had little interest in art; he was moderately versed in history and read some fiction—Aldous Huxley, H. G. Wells and other authors popular in the twenties and early thirties, often suggested by Rasetti. These readings familiarized him with the English language and probably influenced his outlook in the direction of religious agnosticism and of a liberalism rooted in Fabian ideas. He did not accept them without criticism—in fact he often contrasted them with idealistic, semifascist doctrines of which he was equally critical. The latter he had absorbed from the Italian atmosphere and learned from persons such as philosopher Giovanni Gentile while collaborating at the *Enciclopedia italiana*. In American politics he was slightly right of center. He said that he was Republican, but I know for sure that he did not hesitate to vote Democrat if he thought that the Democrats had a better candidate. Living among young scientists prevalently very liberal in outlook, he tended to stress his conservatism to stimulate discussion and to make them analyze the reasons for their convictions. In general he avoided "high-brow" activities but constantly used his original mind and analytical powers to form an independent opinion on every subject he considered.

Fermi was not a promoter of large enterprises, but he infused extraordinary enthusiasm in everybody involved with him in scientific matters. Everybody, from technicians or mechanics to top administrators, rapidly gained immense confidence in his ability and scientific judgment. Above all, his immediate collaborators found that they usually surpassed themselves whenever they worked with him. He went to great pains to instruct them and often gave special lectures on the problems being studied. I often thought of him as a superb conductor who spurs his musicians on by revealing fine points of interpretation and execution to them, and ultimately obtains a uniquely superlative performance.

Another important part of Fermi's makeup, one which was necessary to his success, was his unusual physical strength. He was not an athlete in the technical sense; although he played tennis, swam,

The War Years

skied, hiked in the mountains, and fished, he was not very proficient at any of these sports. He had never taken lessons or bothered to improve his techniques, but his endurance was extraordinary and sometimes allowed him to win competitions against skilled adversaries by the simple method of wearing them down. In the hottest hours of a humid New York summer he would challenge younger physicists to play tennis under the broiling sun, and after an hour or so would comment on their lack of vigor. He showed the same stamina in his scientific work; for example, he thought nothing of spending several hours on arduous calculations, perhaps using the time from five to seven every morning, before his regular, intense working day.

Initially the Fermis stayed for a few weeks at the King's Crown Hotel near Columbia University; then they moved to an apartment in the same neighborhood. Finally they bought a house in Leonia, New Jersey, a town where the chemist Harold Urey and some other professors had their homes.

Once settled in America, Fermi increased his efforts to improve his spoken English and to assimilate American culture. These efforts had begun on previous visits. One summer in Ann Arbor, to improve his pronunciation of the letter r, he had spent hours with a patient and kind young physicist, repeating "Rear Admiral Byrd wrote a report concerning his travels in the southern part of the earth." His success in learning English pronunciation was modest, however, and he retained a strong Italian accent. This was a subject on which he was sensitive. He purposely studied contemporary Americana and read the comic strips,[2] but he did not try to follow baseball (and Laura Fermi says that he found excuses for avoiding gardening). Among adult immigrants, I have never seen a comparably earnest effort toward Americanization. He became an American citizen at the earliest date permitted by law, July 1944.

At Columbia Fermi found several professional and personal friends. George B. Pegram, head of the Physics Department, was an important one. Pegram, born in North Carolina in 1876, had studied at Trinity College, which later became Duke University, and at Columbia University, where he received his Ph.D. in 1903. He was in Berlin and at Cambridge University as a postdoctoral fellow and returned to Columbia where he remained for the rest of his long and distinguished career. Pegram had been instrumental in organiz-

The War Years

ing the American Physical Society. A kind and true gentleman who inspired confidence in everyone who knew him, he became a trusted advisor of many individuals, of Columbia University, and of the United States government. He had worked in nuclear physics himself, and in 1923 he was one of the scientists on the committee that presented a gift of radium to Mme Marie Curie on behalf of the American people. Personally acquainted with Fermi, Rasetti, Amaldi, and me, he had shown great comprehension of human and political problems, and his manners, outlook, kindness, and readiness to help appealed very much to Fermi. Furthermore, over a period of time, in an unobstrusive and tactful way, Pegram had helped Fermi become acquainted with America, and Fermi appreciated this. Columbia's department of physics was growing rapidly in stature under Pegram's leadership and, aware of the importance of the current developments in physics, he had set his heart on building it up to a position of preeminence. Work on molecular beams under the guidance of I. I. Rabi, one of Pegram's protégés, had already acquired international prominence, and more recently Columbia had entered the neutron field in which John R. Dunning, a young physicist especially well versed in electronics and instrumentation, Dana P. Mitchell, and Pegram himself, had obtained significant results. In addition to the staff, several graduate students and faculty members of other New York institutions did research work at Columbia. The chemistry department had as a member Harold C. Urey, of heavy hydrogen fame (a very lively person who was open to any interesting problem), whose particular expertise was physical chemistry and isotope separation, subjects that would soon be of great practical importance. At Columbia Fermi started immediately to teach with his usual vigor and success. It is noteworthy that in addition to standard courses he gave a special course in geophysics—one of his favorite subjects—in which he used in a simple and straightforward way his deep knowledge of physics to explain the essentials of many complicated geophysical facts, but although he was a brilliant teacher, his research activity had priority.

The news of the discovery of fission reached Fermi shortly after he landed in New York. On January 16 Niels Bohr arrived in America and went to Princeton, where, in a lecture, he announced the discovery of fission and its interpretation. The young physicist Willis Lamb, later to become famous for his studies on the "Lamb shift"

The War Years

in the hydrogen atom energy levels, attended the lecture and brought the information to Fermi. This, at any rate, is Fermi's version of the events.[3] Herbert L. Anderson, then a graduate student working under Professor Dunning, has a slightly different one.[4] Be this as it may, on January 25 Anderson and Dunning, using an ionization chamber and an oscilloscope, saw the pulses produced by fission, as Frisch had seen them a few days earlier in Copenhagen. The next day Bohr and Fermi attended a conference on theoretical physics in Washington. Fermi had been informed of the Columbia experiment by telegram, and he and Bohr discussed the phenomenology of fission.

If a uranium nucleus splits into two large fragments, each fragment will have more neutrons than a stable nucleus of the same atomic number as the fragments. The fragments can then rid themselves of the excess neutrons either by instantaneous evaporation of the neutrons or by successive beta emissions. (In fact both phenomena occur.) The neutrons that are instantaneously emitted may cause further fissions in neighboring uranium atoms, and if there are enough secondary neutrons and enough surrounding uranium, a chain reaction may start.

These ideas are fairly simple, and they occurred independently to many physicists, but how could one translate these qualitative speculations into solid reality? And what about the possible technological consequences of a chain reaction?

One man who was keenly aware of the potential applications of nuclear physics and who was destined to play an important part in their development was Leo Szilard,[5] a Hungarian physicist who then resided in New York. He had been one of Planck's very few students, and his Berlin dissertation on thermodynamics and information theory had opened a new avenue of research. When Germany came under Nazi domination, he moved to England, where he worked in nuclear physics, particularly with neutrons. He barely missed discovering slow neutrons but made other important contributions to nuclear physics (for instance, the Szilard-Chalmers method for separating radioactivities). Unconventional and unwilling to accommodate himself to university routine and schedules, Szilard liked to work as an individual in complete freedom. Besides physics he was deeply interested in biology, economics, and politics; he had a brilliant mind and strong political convictions, and he followed world events closely, determined to influence them if at all possible. He cultivated an air of mystery and enjoyed meeting important persons outside the world of science, especially businessmen and politicians. This was not easy for a foreigner to do in a country as insular as the United States was in 1939. The urgency, magnitude, and importance of the problems offered by the possible applications of nuclear physics, as Szilard described them, were not easily grasped by most Americans, perhaps because he indulged himself in sketching seemingly fantastic schemes (which, however, were often closer to reality than "practical" men deemed possible). Szilard took out many patents and by about 1934 had already patented a strange composition of matter in which many substances, some of them radioactive, were mixed together to produce neutrons and secondary reactions. I do not believe that this patent had special value, but it shows his alertness, his interest in practical applications, and his faith in the future of atomic energy. Szilard resided at the King's Crown Hotel and often went to visit the physicists at Columbia University, although he did not have a regular position there. He was also a friend of the other émigré Hungarian scientists, John von Neumann,[6] Eugene Wigner, and Edward Teller, who were of about the same age and had similar backgrounds. Some of them had been schoolmates in Hungary, and now in the United States they were among the most politically active foreign scientists. In addition to these scientists Szilard also knew a number of important persons outside academic circles, including Lewis Strauss. All these men were to play important parts in the development of atomic energy.

In the early months of 1939 the fundamental experiments on fission were repeated and improved upon in many laboratories throughout the world. Secondary neutrons were sought and found, and delayed neutrons also were discovered. This work, however, was not highly quantitative, as one would expect from the speed at which it was done. The study of the chain reaction was taken up independently by von Halban, Joliot, and Kowarski in France, by Fermi, Szilard, and others in the United States, and by other groups in other countries; but no important exchange of information took place among the groups, in part because of self-imposed censorship. Szilard and Fermi were extremely different in personality, habits of work, outlook on life, and almost everything else; they had high regard for each other, but could scarcely work together on the

The War Years

same experiment. They communicated their results to each other but collaborated in only one experiment. Anderson, who participated in the Columbia work, describes the nature of the differences between Fermi and Szilard as follows:

> Szilard's way of working on an experiment did not appeal to Fermi. Szilard was not willing to do his share of experimental work, either in the preparation or in the conduct of the measurements. He hired an assistant to do what we would have required of him. The assistant, S.E. Krewer, was quite competent, so we could not complain on this score, but the scheme did not conform with Fermi's idea of how a joint experiment should be carried out, with all work distributed more or less equally and each willing and able to do whatever fell to his lot. Fermi's vigor and energy always made it possible for him to contribute somewhat more than his share, so that any dragging of feet on the part of the others stood out the more sharply in contrast.[7]

I will report primarily the work of Fermi's group, although one should remember that it was not isolated and that there was continuous exchange of information with other scientists at Columbia, Princeton, and a few other American universities, where different problems related to nuclear fission were under study. Fermi soon concentrated his efforts and those of his coworkers—H. L. Anderson, E. T. Booth, J. R. Dunning, G. N. Glasoe, and F. G. Slack—on obtaining quantitative information on secondary neutrons.

Fermi described the work at Columbia University in his last address before the American Physical Society, delivered informally and without notes at Columbia's McMillan Theater on January 30, 1954.

Fermi's account supplies a great amount of technical and historical information; in particular, it shows that most of the new ideas were developed during the Columbia period. The use of graphite as a moderator rather than hydrogen, the most obvious material, occurred independently to Pegram, Szilard, Fermi, and Placzek.[8] Fermi invented the trick of counteracting the affect of resonance absorption by lumping the material. The effect of impurities was understood, as well as the method of measuring them by exponential experiments. Thus several crucial advances in reactor physics were accomplished during the first two years. Basic patents on the "neutronic reactor" were granted, after the end of the war, to Fermi and various collaborators. They assigned these patents to the government without compensation.[9]

It was natural for Fermi to use the normal uranium-graphite

The War Years

approach rather than the isotope-separation approach, which was the other alternative for making a reactor. The problems of separating isotopes on a large scale were enormous, and at that time they may have seemed insurmountable to Fermi. He felt confident that he could handle the neutron physics needed in the natural uranium-graphite approach and work it out in the minutest detail. He would make a reactor, if at all possible, by effectively using the very last neutron available.

Fermi's 1954 speech implicitly reveals why he was, at least in the beginning, the center of the reactor work: He was without doubt the greatest living expert on neutrons, and he had a rare combination of experimental and theoretical talents that was perfectly suited for the impending job. Furthermore, his personality attracted able coworkers; and finally, he had unbounded energy and physical stamina. Fermi's intuitive knowledge of neutron behavior was in some ways similar to that of an expert on the behavior of radio circuits. He did not need lengthy calculations to predict the results of an experiment on the diffusion of neutrons; his guess was almost always very close to the truth. Nevertheless, he always calculated the results that could be expected and compared his approximate and simple theories with the experimental results. Thus he accumulated a great store of data, which he preserved and ordered in a meticulous way so as to have them available when needed. He called the file his "thesaurus," and it was indeed a treasure of data, numbers, and formulas ready for immediate use. As time went on, the thesaurus could no longer be contained in a large envelope, but required one or two filing cabinets. Because of his unequaled experience with neutrons, Fermi acquired the reputation of being an infallible oracle. When, in the early days of the construction of large piles, reactor engineers were stumped by the lack of nuclear data, they would put their problems to Fermi, who would protest that he could not help them because the number they wanted, usually a nuclear cross section, had not been measured and could not be predicted. One of the engineers said that the regular procedure at that point was to ignore Fermi's protests and to recite slowly a series of numbers while watching his eyes closely; the correct number would produce an involuntary twinkle in his eyes.

One of the reasons why a large proportion of the physicists involved in America's earliest efforts to release nuclear energy were

The War Years

Europeans is that in 1939 and 1940 the development of radar had first priority among leading scientific administrators of the United States.[10] For several years the Massachusetts Institute of Technology's Lincoln Laboratories, where work was being done on radar, absorbed scientists who only later could go to work on the release of nuclear energy. Very correctly, radar was deemed more immediately useful for war purposes and was given precedence over atomic nuclei. Indeed, the atomic project was considered rather visionary even by scientists who later became its proponents. Consequently, it was left in the hands of people who could not be employed in the radar center of the MIT laboratories because they were aliens, or for other reasons. A. H. Compton, in discussing this subject, says that the European scientists were more alert to the Nazi danger than American scientists and thus were anxious to be ahead of Germany in the making of any decisive weapon. He adds that American scientists were better informed on the relationship between science and government in the United States and therefore relied on private sources for funding atomic research rather than seeking help from the federal government, and finally he says that, in effect, "the appointment of the government's Advisory Committee on Uranium (by President Roosevelt in 1939) was to retard rather than to advance the development of American uranium research."[11]

To understand Fermi's role in the history of the uranium project, some background information is needed. I will give the necessary minimum; the interested reader can find a more complete account in the official sources.[12]

Leo Szilard, always sensitive to the implications of atomic energy, had immediately seen the great practical importance of fission as a source of energy and perhaps as an explosive. While making experiments to gain a clear idea of its possibilities, he had alerted Lewis Strauss to some of these as early as January 25, 1939; afterward he kept Strauss posted on developments.[13] Fermi and Pegram, on the other hand, thought that the government should be alerted, and on March 16, Pegram wrote to Admiral Hooper, technical assistant to the Chief of Naval Operations, that he and Fermi would be glad to give him information on a possible new explosive. His cautiously worded letter is in part reproduced on the following page.

The War Years

March 16, 1939

Admiral S. C. Hooper
Office of Chief of Naval Operations
Navy Department
Washington, D.C.
Dear Sir:

... Experiments in the physics laboratories at Columbia University reveal that conditions may be found under which the chemical element uranium may be able to liberate its large excess of atomic energy, and that this might mean the possibility that uranium might be used as an explosive that would liberate a million times as much energy per pound as any known explosive. My own feeling is that the probabilities are against this, but my colleagues and I think that the bare possibility should not be disregarded, and I therefore telephoned ... this morning chiefly to arrange a channel through which the results of our experiments might, if the occasion should arise, be transmitted to the proper authorities in the United States Navy.

Professor Enrico Fermi who, together with Dr. Szilard, Dr. Zinn, Mr. Anderson, and others, has been working on this problem in our laboratories, went to Washington this afternoon to lecture before the Philosophical Society in Washington this evening and will be in Washington tomorrow. He will telephone your office, and if you wish to see him will be glad to tell you more definitely what the state of the knowledge on this subject is at present.

Professor Fermi ... is Professor of Physics at Columbia University ... was awarded the Nobel Prize. ... There is no man more competent in this field of nuclear physics than Professor Fermi. ...

Professor Fermi has recently arrived to stay permanently in this country and will become an American citizen in due course. ...

Sincerely yours,
George B. Pegram
Professor of Physics

As a result of the letter and the various telephone calls, two days later, on March 18, Fermi lectured at the Navy Department to a group that included naval technical experts and two civilian scientists. The minutes, written by Captain G. L. Schuyler (later the Chief of Naval Ordnance), show that Fermi gave a fairly complete description of things to come, but in very conservative terms. Nevertheless, it seems that Fermi's presentation was understood and appreciated, and Ross Gunn, a young physicist at the Naval Research Laboratory, was so impressed by what he heard that he wrote a report to his chief, Admiral Bowen, who allocated Columbia $1500 to help in the investigations. Although the sum was puny, it indicated goodwill.

The War Years

The various groups at Columbia continued their work with the ordinary university support. Their main finding, reported in a paper by Anderson, Szilard, and Fermi,[14] was that the presence of uranium in a tank filled with a manganese sulfate solution increased the radioactivity produced by a neutron source. This was a crucial experiment, because it showed that more neutrons were produced than were absorbed by the uranium, a necessary condition for a chain reaction. Anderson, in commenting on this work, adds:

> When an attempt was made to deduce what the ratio was of fast neutrons emitted per thermal neutron absorbed by uranium, it became apparent that there was a large correction due to the resonance absorption by the uranium. Our discussion as to what to do about this proceeded aimlessly for a while. Then Fermi asked to be left alone for 20 minutes. This was long enough for him to make a rough estimate of this effect, and this was duly recorded in the paper. He never revealed, neither to Szilard nor to me, the details of how he arrived at this estimate, presumably because his basis was largely intuitive. Fermi was never far wrong in such things, and he was taken at his word. The episode did pinpoint the importance of the resonance absorption and provided the clue of how, by lumping the uranium, the losses due to the resonance absorption could be reduced. It also became clear that thermal neutron absorption by hydrogen was too large for water to be a usable medium for slowing down neutrons in a chain reaction.[15]

During the summer of 1939 Fermi went to his beloved Ann Arbor to do theoretical work on muons, a subject very remote from the occupations of the previous winter. Szilard, in the meantime, impatient and frustrated by his inability to obtain backing for the nuclear chain reaction enterprise, decided to approach the President of the United States directly. To do this he first had to have a suitable channel of communication and a scientific authority who would command attention. The intermediary with access to President Roosevelt was Alexander Sachs, an economist who worked with the Lehman Corporation, and the scientific authority was nobody less than Albert Einstein, the greatest living physicist. Wigner and Szilard discussed the letter they would first submit to Einstein, conferred with Sachs, and then took a draft of the letter to Einstein at his summer house in Long Island. The letter was discussed, signed by Einstein, and given to Sachs for transmittal to the president. It read as follows:

The War Years

Albert Einstein
Old Grove Road
Peconic, Long Island
August 2nd, 1939

F. D. Roosevelt
President of the United States
White House
Sir:

Some recent work by E. Fermi and L. Szilard, which has been communicated to me in manuscript, leads me to expect that the element uranium may be turned into a new and important source of energy in the immediate future. Certain aspects of the situation which has arisen seem to call for watchfulness and, if necessary, quick action on the part of the Administration. I believe therefore that it is my duty to bring to your attention the following facts and recommendation.

In the course of the last four months it has been made probable—through the work of Joliot in France as well as Fermi and Szilard in America—that it may become possible to set up a nuclear chain reaction in a large mass of uranium, by which vast amounts of power and large quantities of new radium-like elements would be generated. Now it appears almost certain that this could be achieved in the immediate future.

This new phenomenon would also lead to the construction of bombs, and it is conceivable—though much less certain—that extremely powerful bombs of a new type may thus be constructed. A single bomb of this type, carried by boat and exploded in a port, might very well destroy the whole port together with some of the surrounding territory. However, such bombs might very well prove to be too heavy for transportation by air.

The United States has only very poor ores of uranium in moderate quantities. There is some good ore in Canada and the former Czechoslovakia, while the most important source of uranium is Belgian Congo.

In view of this situation you may think it desirable to have some permanent contact maintained between the Administration and the group of physicists working on chain reactors in America. One possible way of achieving this might be for you to entrust with this task a person who has your confidence and who could perhaps serve in an unofficial capacity. His task might comprise the following:

a. to approach Government Departments, keep them informed of the further development, and put forward recommendations for Government action, giving particular attention to the problem of securing a supply of uranium ore for the United States.

b. to speed up the experimental work, which is at present being carried on within the limits of the budgets of University laboratories, by providing funds, if such funds be required, through his contacts with private persons who are willing to make contributions for this cause, and perhaps also by obtaining the co-operation of industrial laboratories which have the necessary equipment.

The War Years

I understand that Germany has actually stopped the sale of uranium from the Czechoslovakian mines which she has taken over. That she should have taken such early action might perhaps be understood on the ground that the son of the German Under-Secretary of State, von Weizsäcker, is attached to the Kaiser-Wilhelm-Institut in Berlin where some of the American work on uranium is now being repeated.

Yours very truly,
Albert Einstein

A few days after the writing of this letter, on September 1, 1939, Hitler invaded Poland, and the Second World War began.

The letter could not be delivered until October 11, when Sachs met with Roosevelt, presidential assistant General E. M. ("Pa") Watson, and two ordnance experts. At the end of the interview the president remarked to Sachs: "Alex, what you are after is to see that the Nazis don't blow us up."[16] Then he turned to Watson and said: "This requires action." Roosevelt appointed a Committee on Uranium, composed of Lyman J. Briggs, director of the National Bureau of Standards and as such the official physicist of the government; Commander Hoover and Colonel Adamson, the two ordnance experts; and Sachs himself. Briggs recruited additional members including the three Hungarian physicists, Szilard, Wigner, and Teller. The committee obtained $6,000 from the army and navy, invited several other physicists, including Fermi, to its meetings, and kept the president informed of the progress of the uranium work. In June 1940 at a meeting attended by Briggs, Urey, Merle A. Tuve, Wigner, Gregory Breit, Fermi, Szilard, and Pegram, the committee recommended an expenditure of $140,000 for further study of the nuclear properties of uranium and for procurement of graphite and metallic uranium of about one-fifth the minimum amount estimated necessary for a chain reaction.

At this time, however, under the pressure of events in Europe—the passage, with Hitler's invasions, from the "phony" to the real war—the research effort of the United States was being reorganized. The possibility of America's involvement in the war was becoming apparent to informed people if not to the whole citizenry.

Vannevar Bush, a professor of electrical engineering, an inventor, and at that time president of the Carnegie Institution, had been president of the National Advisory Committee for Aeronautics (NACA) and had experience in organized research. Bush, the prime mover of the reorganization, consulted with other scientists and adminis-

The War Years

trators (including K. T. Compton, a physicist at Massachusetts Institute of Technology; J. B. Conant, a chemist and the president of Harvard University; and F. B. Jewett, president of the National Academy of Sciences) and persuaded President Roosevelt to organize a National Defense Research Committee (NDRC) to foster defense activities, centralize the efforts, and assist the various initiatives then in progress—in short, to mobilize science for war. Bush was appointed president of the NDRC, and the uranium committee was placed under its jurisdiction.

In the early autumn of 1940 Bush reorganized the uranium committee. Briggs was retained as chairman; the military members were dropped; foreign-born scientists were excluded under the assumption that this would improve security; and Tuve, Pegram, Beams, Gunn, and Urey were added to the membership. Bush also restricted the publication of uranium papers for reasons of security, but this was only official implementation of a policy that had been initiated by the scientists themselves, primarily at the urging of Szilard and Breit. Most important, the NDRC provided the funds that had been requested by the uranium committee. The complex history of the successive changes and the organization of numerous committees, of the enlistment of great numbers of American scientists, of the interplay between the scientists, the military, and the huge industrial companies, which eventually led to the construction of the atomic bomb, has been treated elsewhere exhaustively and with documentation.[17] In these pages I shall only mention briefly whatever is necessary to explain Fermi's activities.

Fermi took little interest in organizational problems and was completely absorbed in the effort to achieve a chain reaction. He would attend meetings and, to the best of his ability, help in administration, but he did so without relish; and, of course, he was sensitive about being foreign-born and an alien. All in all, he kept himself far removed from the political activism of Szilard and Wigner and concentrated primarily on technical problems. It was nevertheless apparent to the American administrators that Fermi was the key man technically; they admired him as a scientist and necessarily brought him into their councils for advice.

During the spring of 1940 Fermi lectured for a few weeks at the University of California at Berkeley as a Hitchcock Professor, and

The War Years

for the first time since 1938 we saw each other. Needless to say, I was happy to see a dear friend of old standing. I, too, was then working on fission, and we talked at great length about physics; but because of the policy of voluntary secrecy, Fermi did not mention the progress that had been made at Columbia beyond what I had read in the published literature. We discussed in detail his work on the passage of muons through matter and even tried an experiment on this subject which was not successful. Fermi also saw E. O. Lawrence[18] at Berkeley, but I doubt that they had serious scientific conversations. The 60-inch cyclotron had recently been activated, and Fermi and I used it to produce fission in uranium, using alpha particles as projectiles.[19]

After he returned to Columbia, Fermi pursued the experiments on the absorption of neutrons in graphite. The use of pure graphite as a moderator, of normal uranium as a fuel, and the ingenious device of lumping the uranium rather than mixing it uniformly with the moderator were now firmly established requirements for achieving a chain reaction without isotope separation. The theoretical analysis, an extension of the Italian work on the motion of neutrons in hydrogenous substances, was taking form, and by the spring of 1941 Fermi was instructing his collaborators on the essentials of the theory of the chain reaction, including the famous four-factor formula, $k = \Sigma \eta p f$, which connects the reproduction factor, k, with the fast neutron multiplication, Σ; the number of neutrons emitted per captured slow neutron, η, the resonance escape probability, p, and the fraction of neutrons absorbed in fissionable material, f.

Much of the standard nomenclature in nuclear science was developed at this time. For instance, a nuclear reactor was called a *pile*. I thought for a while that this term was used to refer to a source of nuclear energy in analogy with Volta's use of the Italian term *pila* to denote his own great invention of a source of electrical energy. I was disillusioned by Fermi himself, who told me that he simply used the common English word *pile* as synonymous with *heap*. To my surprise, Fermi seemed never to have thought of the relationship between his *pile* and Volta's. The word *barn*, denoting the unit of area 10^{-24} cm^2 used for measuring nuclear cross sections, simply reflected the relative size of the cross sections: they were, in American slang, "as big as a barn door." This is why a drawing of the barn appears on the cover of the compilations of the cross sections from multiplied.

Fermi at the Age of Sixteen

Ostia, 1927. Segrè, Persico, and Fermi

The Rome Institute of Physics on Via Panisferna

The Library of the Rome Institute

Rome, 1931. Fermi, Corbino, Trabacchi, Sommerfeld, and Zanchi

hundertmal kleiner sind wie die der erlaubten Übergänge.

7. **Die Masse des Neutrinos.** Durch die Übergangswahrscheinlichkeit (32) ist die Form des kontinuierlichen β-Spektrums bestimmt. Wir wollen zuerst diskutieren, wie diese Form von der Ruhmasse μ des Neutrinos abhängt, um von einem Vergleich mit den empyrischen Kurven diese Konstante zu bestimmen. Die Masse μ ist in dem Faktor p_σ^2/v_σ enthalten. Die Abhängigkeit der Form der Energieverteilungskurve von μ ist am ausgeprägtesten in der Nähe des Endpunkts der Verteilungskurve. Ist E_0 die Grenzenergie der β-Strahlen, so sieht man ohne Schwierigkeit dass die Verteilungskurve der β-Strahl Energie für Energien E in der Nahe von E_0 proportional ist durch zu

$$(36) \quad \frac{p_\sigma^2}{v_\sigma} = \frac{1}{c^3}(\mu c^2 + E_0 - E) \cdot \sqrt{(E_0-E)^2 + 2\mu c^2(E_0-E)}$$

Manuscript of the Beta Ray Theory (*FP* 80*b*)

Solvay Conference, 1933.
Standing: E. Henriot, F. Perrin, F. Joliot, W. Heisenberg, H. A. Kramers, E. Stahel, E. Fermi, E. T. S. Walton, P. A. M. Dirac, P. Debye, N. F. Mott, B. Cabrera, G. Gamow, W. Bothe, P. Blackett, M. S. Rosenblum, J. Errera, E. Bauer, W. Pauli, J. E. Verschaffelt, M. Cosyns, E. Herzen, J. D. Cockcroft, C. D. Ellis, R. Peierls, A. Piccard, E. O. Lawrence, L. Rosenfeld. *Seated:* E. Schrödinger, I. Joliot, N. Bohr, A. Joffé, Mme. Curie, O. W. Richardson, P. Langevin, E. Rutherford, T. de Donder, M. de Broglie, L. de Broglie, L. Meitner, J. Chadwick.
Absent: A. Einstein, C. E. Guye

(Middle left): Summer 1934. D'Agostino, Segrè, Amaldi, Rasetti, and Fermi

The "Roman Sign": Ionization Chamber Used for Measuring Radioactivity. It was simple and accurate, and we built similar ones wherever we went.

Landing in New York:
The Founding of the American Branch of the Fermi Family.
Laura, Giulio, Nella, Enrico

Fermi and Robert Oppenheimer

Los Alamos, Summer 1945. Lawrence, Fermi, and Rabi

Fourth Anniversary Reunion of Some of the Participants in First Pile Experiment, 2 December 1946. *Top Row:* N. Hilberry, S. Allison, T. Brill, R. Nobles, W. Nyer, M. Wilkening. *Middle Row:* H. Agnew, W. Sturm, H. Lichtenberger, L. W. Marshall, L. Szilard. *Front Row:* E. Fermi, W. Zinn, A. Wattenberg, H. Anderson

Target Holder for Cyclotron, Fermi's Own Handiwork

Fermi Lecturing

The War Years

Brookhaven National Laboratory. I think that the word *barn* was coined in this context by Bethe. Examples of this kind could be

The only purpose of NDRC was to mobilize science for war. It was interested in producing a chain reaction solely for its possible military applications. The development of a super-explosive was paramount; the production of nuclear power for submarine propulsion and other uses was, at best, secondary. Because it was now apparent that ordinary uranium was inadequate as an explosive, the only possibility for a bomb at that time lay in the separation of the rare uranium isotope 235, a process deemed to be of forbidding difficulty by most physicists, including Fermi. There was an almost dogmatic acceptance of the idea that isotopes were practically inseparable except by the mass spectrograph; and especially with uranium, whose mass is very large compared with the mass difference of the isotopes, the prospects appeared bleak. That the separation was ultimately accomplished, and on a large scale, was due to American science and technology, and to such men as Lawrence, Urey, Dunning, E. T. Booth, and K. Cohen (Fermi hardly participated in this enterprise), and to such industrial companies as Du Pont, Union Carbide, and Tennessee Eastman.

There was, however, an alternative to isotope separation: uranium 238, as had been demonstrated by Hahn and Meitner, could capture neutrons, becoming a beta emitter with a half-life of about 23 minutes. The decay product of this radioactivity had to be a new element of atomic number 93 and mass 239. In the summer of 1940 E. McMillan and P. A. Abelson, at Berkeley, identified this substance, the first transuranic element discovered, later called neptunium (Np). Np^{239} decays by beta emission with a period of 2.3 days, thus giving rise to an isotope of element 94 of mass 239, later called plutonium (Pu). It was probable that such a nucleus would undergo fission when it was bombarded with slow neutrons. This hypothesis was based on the systematics of the binding energy and was corroborated by the knowledge that in uranium the isotope of odd mass 235 undergoes slow neutron fission whereas the even isotope U^{238} can be fissioned only by fast neutrons. If a chain reaction with natural uranium could be accomplished, a considerable amount of Pu^{239} would be formed in the reactor and could be chemically separated. This procedure, though very difficult, would obviate separating the isotopes and would provide a new material for a

The War Years

possible bomb. These ideas had been in the air, so to speak; and they sprung up at several places, but wartime secrecy forbade their promulgation.

In December of 1940[20] I visited Fermi at his home in New Jersey, and this time he was willing to discuss at least part of his current work. It was clear that if Pu^{239} had a half-life long enough to make separation and accumulation practical, and if it underwent slow neutron fission, it would be a nuclear explosive. A controlled chain reaction with natural uranium would then acquire new importance because it would be the source of material for a nuclear bomb.

On a long, cold walk along the Hudson River, which was covered by ice floats, we developed this subject at length. The Radiation Laboratory at Berkeley had a powerful neutron source in the newly built 60-inch cyclotron that might form enough Pu^{239} that its properties could be studied. Since I was returning soon to Berkeley and had experience in the required techniques, Fermi asked me to undertake the investigation. An unusually long bombardment period was needed, and I said that we should discuss the project with E. O. Lawrence, the director of the laboratory. Fortunately, Lawrence was in New York, and on December 16, 1940, we arranged a meeting of Fermi, Lawrence, Pegram, and myself (at Columbia University) at which we discussed the details of the irradiation. I returned to Berkeley and, at the beginning of 1941, tried to start the work, but the undertaking was beyond the power of one man. I asked the help of two friends, Glenn T. Seaborg and Joseph W. Kennedy, with whom I had worked on various problems in radiochemistry. It turned out that Seaborg and Kennedy had already done work on the products of deuterium bombardment of uranium, some of which are isotopes of neptunium and plutonium, but at the time no method was known for chemically separating neptunium from plutonium. It was important that we measure the fission cross section of isotope plutonium 239 because this was the isotope that would be most abundantly formed in a reactor if such a device became a reality. We therefore concentrated on obtaining, by neutron bombardment of uranium, a sample of neptunium 239 that, without further separation, would contain the desired plutonium 239 formed by decay. We prepared such a sample, examined it for fission, and obtained a positive result; but the sample was not suitable for a quantitative

The War Years

measurement of the fission cross section because the separation of neptunium was made with several milligrams of carrier, and our sample was too thick. Soon, however, a chemical separation of neptunium and plutonium was developed by Kennedy, Seaborg, and Arthur Wahl, and our sample was thinned to make quantitative measurements possible. By direct comparison with U^{235} we measured the ratio of the slow neutron fission cross sections of the two isotopes. A few months later we made similar measurements, using fast neutrons. Our work showed that plutonium 239 was a nuclear explosive and that there were now two ways of making the bomb: by the separation of uranium isotopes or by the preparation of plutonium 239 in quantity by means of a nuclear reactor, if such a device was perfected. I could not communicate these results to Fermi because we were both aliens and were discouraged from communicating directly. The experiments, consequently, were reported in secret letters from Seaborg to Briggs, for transmittal to Fermi.[21]

In the spring of 1941 there were signs that important American scientists were becoming more interested in atomic energy. Whereas previously Europeans—most of them recent immigrants—had had the preponderant voice, A. H. Compton and E. O. Lawrence (in addition to those mentioned earlier) began giving greater attention to atomic energy and worried that the rate of progress was too slow. They brought to the enterprise organizational talent, administrative experience, and a more realistic view of the scale of the expenses and manpower requirements. The exchange of information with British scientists also helped the uranium project considerably. In May 1941, in a report to the National Academy of Sciences, Lawrence outlined the possibility of using element 94 for reactors and bombs.[22] The increased interest in the uranium work, and the recognition that an organization comprising the National Academy, NDRC, and other agencies was not well adapted to the situation stimulated Bush to a reorganization that would remedy the defects. He persuaded President Roosevelt to establish the Office of Scientific Research and Development (OSRD), which the President did by an executive order promulgated June 28, 1941. OSRD was to have easy access to the President and was to serve as a center for mobilizing the scientific resources of the nation and applying them to defense problems. Bush became director of OSRD. Conant re-

The War Years

placed him as chairman of the NDRC, and the Committee on Uranium became OSRD Section One, or S1.

At the beginning of December 1941, just before the Japanese attack on Pearl Harbor and America's entry into the war, the technical situation was this: no chain reaction had been achieved; no appreciable amount of uranium 235 had been separated; and only minute amounts of plutonium 239 had been produced. The nuclear properties of these substances were fairly well known, however, and studies had been made of the engineering problems in their large-scale production; moreover, the critical size of the bomb almost certainly was within practical limits. The psychological change of the last year, however, was perhaps the most important development. In the words of the Smyth report:[23] "Possibly Wigner, Szilard, and Fermi were no more thoroughly convinced that atomic bombs were possible than they had been in 1940, but many other people had become familiar with the idea and its possible consequences. Apparently, the British and the Germans, both grimly at war, thought the problem worth undertaking." At the end of 1941, in this changed atmosphere, it was decided to proceed with colossal expenditures of effort and money, which would have been considered preposterous even a few months earlier.

An important step toward this decision had been the convening, in November 1941, of a special review committee of the National Academy of Sciences to make definite recommendations to NDRC. Well before the committee met, A. H. Compton visited Fermi at Columbia to gather firsthand information on the feasibility of a nuclear bomb. Compton had a certain diffidence for a "recently arrived émigré," but Fermi's technical explanations were clear and to the point. Compton was able to follow his calculations and later to reconstruct them in his mind. When the occasion arose he fully endorsed them.

The decision to make an all-out effort was announced on December 6, 1941, the day before Pearl Harbor, and had as an immediate consequence an administrative reorganization of the uranium project.

In the reorganization of the uranium project in December 1941 Arthur H. Compton was placed in charge of all scientific work related to the chain reaction and the construction of the pile, a move that almost at once affected Fermi directly. In the middle of Janu-

The War Years

ary 1942 Compton, who was a professor of physics at the University of Chicago, decided to bring all the work and workers under him, including the Columbia group, to Chicago and form a large organization under the name Metallurgical Laboratory.

Fermi's move to Chicago occurred in stages. Part of Fermi's group moved at once to Chicago where, under the direction of Fermi and Anderson, they began to set up a bigger pile on the campus of the university. Fermi, however, did not immediately settle in Chicago. During the winter and part of the spring he commuted between Chicago and New York, where the original experiments to determine absorption and multiplication of neutrons in materials were continued at Columbia, chiefly under the direction of Walter Zinn, who moved to Chicago only after these experiments were completed. The experiment in Chicago was called the "intermediate pile" because it was to be an intermediate step between the Columbia experiment, in which a relatively small amount of material was tested, and the final pile, which had to reach criticality.

Fermi accepted the decision to move the pile work to Chicago without enthusiasm. He was perfectly at ease with his Columbia group (Anderson, Feld, Marshall, Weil, and Zinn), and the work he was doing was highly congenial. It involved a mixture of theory and experimentation best suited to his abilities, and it was on such a scale that he could entirely comprehend it, keep it under control, and affect it with his own hands. In Chicago conditions were different. The Metallurgical Laboratory was becoming a large organization comprising not only the pile work, but also chemistry, metallurgy, engineering, health physics, and other kinds of work; and in this large enterprise the small Columbia group threatened to become lost.

In spite of his strong preferences, however, Fermi had to "change his ways." More and more he had to attend meetings, make reports, give advice on technical matters, and tactfully guide engineers who were entirely new to the problem. Instead of performing the needed experiments as had been the case at Columbia, he now called on competent collaborators to make measurements for him. Although he always reserved for himself the analysis of the data, he regretted how little time he could spend in the laboratory. He told me once that he felt he was doing physics "by telephone."

There was another reason for Fermi's early displeasure with the move to Chicago. After the attack on Pearl Harbor and the subse-

The War Years

quent declaration of war on Italy, he was classified as an enemy alien and became subject to several restrictions. His movements, for instance, were curtailed, and he could not leave New York without special permission—just at the time when he was often needed in Chicago. Nobody could ask for exemption from these restrictions because the reason one would have to give was so secret that it could not be revealed to the authorities administering the regulations against enemy aliens. When the army entered the picture, it was faced with thorny problems, since on the one hand it was hard to stretch security rules to clear enemy aliens, and on the other it would have been foolhardy to exclude a man such as Fermi from the work. Common sense, courage to take risks when necessary, and a profound understanding of security problems made it possible to find workable solutions. At any rate, on Columbus Day 1942, President Roosevelt announced that Italians would no longer be considered enemy aliens. Meanwhile, in the previous winter, Fermi had obtained a permanent permit to travel between New York and Chicago (but not elsewhere).

Fermi's mild irritation about these difficulties was irrelevant; wartime urgency would not permit personal preferences or feelings to interfere with the job. A good judge of the wartime situation, Sir Winston Churchill, said that the alliance of Britain, the United States, and Russia "made final victory certain unless it broke in pieces under the strain, or unless some entirely new instrument of war appeared in German hands."[24] It was the responsibility of the scientists of the Grand Alliance to prevent such a catastrophe.

By May 1942 the transfer of Fermi's group from Columbia to Chicago was completed, and Fermi settled in Chicago, where his family soon joined him. They thought the move would be only temporary, but after the war they remained in Chicago permanently.

The move to Chicago obviated Fermi's travel difficulties, but he faced another annoyance with security: in Chicago his mail, as he could easily tell by its appearance, was being opened before he received it. He resented this action as an invasion of his privacy and a betrayal of the trust that was implicit in the work he was doing. He protested to the authorities, who first denied the fact and then later admitted it, made excuses, and said it was all a misunderstanding and a mistake of some low-level employee. They assured him that censorship of his mail would be stopped immediately. But a few days later Fermi found a card in his mailbox, obviously deposited

The War Years

there by mistake, instructing the post office to submit all his mail to a security officer for censorship before delivery. Now really angry, Fermi returned to the authorities to protest against this way of keeping promises. The poor officer with whom he spoke, probably shaken and befuddled by Fermi's indignation, tried to convince him that the attempt to intercept his mail was the work of a foreign spy— he was glad to be informed of the matter, because he would investigate and probably be able to uncover a dangerous enemy agent. The explanation was so ridiculous that Fermi burst out laughing, and this somehow ended the incident.

While the Metallurgical Laboratory was being organized in Chicago, a group at Berkeley under Lawrence undertook initial work and pilot-plant studies for large-scale electromagnetic separation of the uranium isotopes. During the summer of 1942 another Berkeley group under the leadership of J. Robert Oppenheimer began to study more specifically the possibility of a fast reaction and the feasibility of a bomb. There were also administrative changes: In May 1942, OSRD-S1 was replaced by an executive committee consisting of J. B. Conant (Chairman), L. J. Briggs, A. H. Compton, E. O. Lawrence, E. V. Murphree, H. C. Urey, with H. T. Wenzel and I. Stewart as technical aide and secretary, respectively. This streamlined executive committee was charged with (1) reporting on the program and budget for the next eighteen months for each method of isotope separation; (2) preparing recommendations on how many programs should be continued; and (3) preparing recommendations on what parts of the program should be eliminated.[25]

As the work proceeded in the early months of 1942, the question whether an atomic weapon would be decisive in the present war came sharply into focus, because upon its answer hinged the justification for the all-out effort. In March and June Bush reported to President Roosevelt on the progress of the work. The June report made four major points: (1) It was clear that a mass of several kilograms of Pu^{239} or U^{235} would be explosive and equivalent in effects to several thousands tons of conventional high explosive and, furthermore, that such a bomb could be detonated at will. (2) It was clear that there were four feasible methods of preparing fissionable material: electromagnetic separation of U^{235}, diffusion separation of U^{235}, separation of U^{235} by centrifuge, and production of Pu^{239} by a chain reaction; it was not possible to state definitely that

The War Years

any one of the four methods was superior to the others. (3) It was clear that production plants of considerable size could be designed and built. (4) Given adequate funds and priorities, it was likely that full-scale operation could be started soon enough to be of military importance.[26] The conclusions were reviewed and commented on by Conant, Lawrence, Urey, Compton, Murphree, General Styer (who was representing General G. C. Marshall) and forwarded to the president, who approved them. The report required a commitment of about $100,000,000.

It must be kept in mind that at that time there were still only microscopically small samples of separated isotopes, that the chain reaction had not yet been achieved, and that studies concerning the bomb itself were only rudimentary. To translate the positive laboratory experience into an industrial development was a staggering enterprise. In normal circumstances it would probably have taken decades, but the war urgency spurred the leaders to take risks which would have been deemed preposterous under normal circumstances. The decisions made required very large-scale industrial commitments, the solution of extremely knotty problems of priority in a war-strained economy, and the omission of many traditional steps and procedures in the transition from laboratory experiments to industrial plants. And all this had to be accomplished in absolute secrecy. It is a wonder that such objectives were ever attained. The strain on the individuals, who often found themselves in severe conflict between (on the one hand) their own technical feelings and intellectual beliefs and (on the other) the decisions made for them, was extreme. Scientists were not used to military discipline and industrial ethics, and it was difficult for them to accept the restrictions demanded by the circumstances. One of the unifying forces, however, was their hatred of Hitler and what he stood for.

The army was called upon to take charge of the uranium project in the summer of 1942. As early as the end of 1941, Bush had discussed with the President the advisability of directly involving the military. The possibility of assigning the impending gigantic task to the army, rather than to OSRD, had been mentioned in the March (1942) report, and was restated in June. Action followed, and the Manhattan District of the Corps of Engineers (MED) was formed. After some interim arrangements, on September 17, 1942, Brigadier General Leslie R. Groves was appointed commander of the Manhattan District.[27] Groves, a colonel with considerable construction

The War Years

experience, was given the temporary rank of brigadier general and put in charge of an enterprise of which he had known virtually nothing a few days before. Nevertheless, the choice was a happy one.

Although not an intellectual, General Groves was intelligent, energetic, decisive, and dedicated to his assignment. He had no experience in dealing with scientists (whom he labeled "expensive crackpots" in a good-natured speech that he gave in Los Alamos several months after his appointment), but he soon learned how to treat them, whom to trust, and how to obtain their cooperation for the paramount objective. He had a clear understanding of construction jobs, industrial problems, production schedules, financial matters, and the like, and he knew the industrial world well. He knew how to organize and coordinate an extremely complex enterprise, assigning specific tasks to the various laboratories and delegating the necessary powers. Thus he brought to the project a realistic, practical point of view that balanced the weaknesses of the scientists, who were generally inexperienced in such matters. Unlike the scientists, he was not fascinated by the new vistas which continually opened as the project evolved, and he was not emotionally involved in the technical choices, which he did not always fully understand; but he could keep his eyes constantly on the final goal: obtaining a decisive weapon at the earliest possible date.

As a military man, General Groves had his prejudices just as the scientists had theirs, but he was clear-sighted enough to recognize his limitations even though not always admitting them explicitly, and he had a genuine, if somewhat heavy, sense of humor. Being endowed with integrity, courage, and a willingness to assume responsibilities, he was a source of strength in times of crisis. Groves soon came to have great respect for Fermi, a respect based on objective observation of his performance. The two men got along well together, although they were totally different in temperament, intellectual traits, and background.

General Groves chose as his aides two able officers, General T. Farrel and Colonel K. D. Nichols, and immediately addressed himself to the production and industrial problems, which he easily grasped. His cajoling, pressuring, and patriotic appeal helped bring the big industrial companies into the project and helped establish working relationships between them and the scientists. Groves also labored ably and forcefully to solve the difficult problem of trans-

The War Years

forming these relationships into a cordial collaboration, smoothing rough edges, arbitrating incompatibilities, and explaining to each party the various necessities and points of view.

At the time he took charge of the project, one of the most urgent problems was to build plants for the expanding work. The general rapidly acquired a site at Argonne Forest near Chicago, to house part of the proliferating work of the Metallurgical Laboratory, and a larger site at Oak Ridge, Tennessee. These were followed closely by the sites at Los Alamos, New Mexico, and Hanford, Washington. At once, in early fall 1942, he began construction.

A minor consequence for Fermi of the army takeover is worth recalling. He and a few other key persons in the project were each assigned bodyguards—armed military security men in civilian clothes. Fermi's bodyguard, John Baudino, was a draftee, a lawyer from Illinois of Italian origin. His orders were to accompany Fermi constantly and to protect him. After a short period of uneasiness due to the novelty of the situation, Baudino by his tact, intelligence, and cheerfulness made himself very agreeable to Fermi and to Fermi's friends, who developed a sincere liking for him. He discharged his duties most conscientiously and also learned how to spend his leisure time usefully, performing physics tasks in the laboratory, for instance, or advising scientists how to fill out their income tax returns.

During the months of reorganization and expansion of the project, the group studying the chain reaction at the Metallurgical Laboratory was making great progress. Fermi himself gave a brief, factual description of the work done in Chicago in a paper he delivered before the American Philosophical Society in 1945.[28] H. L. Anderson has given a vivid description of the day-by-day work of the completion of the pile.[29]

By the fall of 1942 the two great production hurdles that stood in the way of a chain reaction were about to be solved. Sufficient quantities of pure graphite and of pure uranium and uranium oxide were being delivered to the physicists. Tests performed during the summer had shown that, given enough of these materials, a natural uranium-graphite reactor would work. We can follow the approach to criticality very closely in the reports of the Metallurgical Laboratory.[30] The struggle to obtain better materials and to make lattices which would give a higher reproduction factor, as described in these reports, is quite dramatic to understanding eyes.

The War Years

In October there was enough uranium and graphite to start building not simply one more exponential experiment but an actual reacting pile, but first the graphite had to be machined and the uranium formed into approximately spherical lumps. By the middle of the month two hundred tons of graphite had been processed. It was apparent then that within a month or two the critical size and a self-sustaining reaction could be attained. The physicists had planned to assemble the pile at the Argonne site, but a strike of labor unions retarded the completion of the building, which should have been finished by October 20. Fermi then proposed to Compton that the pile be assembled under the west stands of Stagg Field, the University of Chicago stadium. This meant that the experiment would be carried out in the heart of the city of Chicago. In Fermi's informed opinion the risk was negligible, and although it worried Compton and General Groves, they decided to let him proceed on the campus of the university rather than wait for the completion of the Argonne building. Actual construction of the pile began in October under Zinn's and Anderson's immediate direction and Fermi's supervision. The physicists worked twenty-four hours a day in two shifts, one of which was headed by Anderson, the other by Zinn. The pile, according to plan, was shaped as a rotational ellipsoid with a polar radius of 309 centimeters and an equatorial radius of 388 centimeters. Most of the uranium lumps had to be placed in the central region for better utilization. The weight of the uranium was approximately 6 tons. To use the material efficiently the purest fuel had to be located more centrally and one had to watch carefully the details of the geometry because they could affect the reproduction factor. The whole structure was supported by a wooden frame. Fermi feared that there might not be enough material to reach criticality, and to reduce parasitic neutron absorption by the nitrogen of the air, he had ordered a huge rubber balloon to enclose the pile. The balloon could be evacuated and air absorption thus eliminated. Ultimately, however, the balloon proved to be unnecessary.

On November 16 Anderson and Zinn began arranging the material according to a general plan, but without a detailed blueprint. Each layer of the pile was built up after discussions with Fermi and extensive calculations by him which took into account the quality of the material available. Cadmium sheets nailed to wooden rods could be inserted in suitable slots to control the pile (cadmium is a strong neutron absorber, and the cadmium sheet acts as a neutron

The War Years

sink, thus reducing the activity of the pile). The movable rods were inserted in the pile and locked in their position with a padlock that could be opened only by keys in possession of Anderson and Zinn. As the work proceeded, the physicists temporarily removed the control rods every day and measured the reactivity of the pile by observing the multiplication of cosmic rays and spontaneous fission neutrons always present in the pile. Volney C. Wilson, who was in charge of instrumentation, made these measurements. From the neutron density found at the observation points it was possible to infer how much material had to be added before a critical size would be reached at which the reaction would become divergent—that is, the neutron density would increase exponentially with time. At the end of November the measurements clearly showed that completion of the fifty-seventh layer of the structure would make the pile critical. This would occur during the night of December 1–2, on Anderson's shift.

Fermi asked Anderson to promise that he would not start the chain reaction in his absence, during the night, and Anderson, true to his word, on completion of the uranium layer did not pull the cadmium control rod. The next morning, December 2, 1942, everything was ready for the final test that should demonstrate the first self-sustaining chain reaction.

About forty persons were present, mostly scientists who had worked on the pile and a few spectators.[31] Fermi had prepared a procedure to approach criticality in a perfectly controlled way, by extracting the control rod little by little and measuring at each step the reactivity of the pile. He could immediately calculate the result of the observation on his slide rule and order the size of the next step. It was thus possible to know exactly at each moment where one stood and to anticipate the result of the next move. As an additional precaution there were emergency control rods that could be inserted by gravity simply by cutting a string, and finally there were volunteers standing on the pile with buckets of a cadmium salt solution ready to dump the solution on the structure in case of unforeseen catastrophe. Needless to say, the extra precautions proved unnecessary. The test proceeded during the morning, and by lunchtime the pile was not yet critical. True to his nature, the imperturbable Fermi called for a lunch recess. (When they were discovered on October 22, 1934, slow neutrons had not been permitted to interfere with lunch). In the early afternoon Fermi completed the morn-

The War Years

ing's work, and after the expected step, at 2:20 P.M., the pile went critical.

The first test was run for twenty-eight minutes at a power not exceeding half a watt, to minimize the production of radioactivity.

After the pile was shut down, Wigner produced a fiasco of Chianti wine that he had bought for the occasion several months earlier. Among those present were Compton and C. H. Greenewalt, a chemical engineer who later became president of the Du Pont Company. He had come for a committee meeting in Chicago at which the participation of the Du Pont Company in the production of plutonium was one of the important topics of discussion. Compton thought that the demonstration of the operation of the pile would impress Greenewalt, as indeed it did. Many years later, however, Greenewalt said[32] that Du Pont had reached a favorable decision on the prospects of the pile before the December experiment and had decided to accept the government contract for the plutonium work. He added that the company was confident that had the experiment failed, a second attempt would have been successful. After the reactor went critical, Compton and Greenewalt left to tell the review committee what they had seen, and then Compton telephoned Conant at Harvard. "Jim," he said, "you will be interested to know that the Italian navigator has just landed in the new world." Then, apologetically, because he had told Conant's S-1 committee that criticality would be reached somewhat later, he added: "The world was not as large as he had estimated and he arrived at the new world sooner than he expected." "Is that so?" Conant asked. "Were the natives friendly?" "Everybody landed safe and happy," Compton assured him.[33]

The experiment of December 2 was a milestone in the development of nuclear energy and is now commemorated by a bronze sculpture by Henry Moore on the site where it occurred. Probably for Fermi, however, the real victory in the making of a natural uranium reactor had been won a few months earlier when he succeeded in building a lattice with $k > 1$, which was tantamount to reaching criticality. In October 1942, while I was in Chicago on a laboratory errand, Fermi had locked me up in a room, alone, to read a few reports on his work. "Read them," he said and left. After an hour or two he returned and found me rather speechless and with bulging eyes. Of course, I knew of the attempts to obtain a chain reaction with natural uranium, but I had no precise idea of how far

The War Years

the work had proceeded, although my own work was dependent on the production of plutonium and hence on a functioning nuclear reactor. The progress reports I read impressed me as though I had seen a critical pile with my own eyes.

I mention this incident because I am sure that for Fermi the December demonstration experiment must have been much less startling than for outsiders. The successful outcome of the experiment was celebrated by a modest party at Fermi's house with his collaborators (without, however, letting Laura Fermi know the reason for the celebration), but there was no letup of the day-to-day work. Too many problems had yet to be overcome.

The Metallurgical Laboratory had an engineering council, chaired by T. V. Moore, to guide the laboratory's studies of a plutonium production pile. Fermi was called upon—several months before the pile went critical—to give technical advice for future plants on such knotty questions as how to cool power reactors. Each of the three competing substances proposed as a coolant—helium gas, water, and liquid bismuth—had advantages and disadvantages, and the decision was deferred to a laboratory policy committee, which finally favored water. Such questions aroused strong emotion among the scientists and could not be resolved by calculation because they continually sprouted complicated ramifications and were in many ways matters of judgment. Fermi tackled this type of problem by personally calculating or estimating everything amenable to analysis. He was always, at least outwardly, rational and objective (however intuitive he may have been inwardly), and his enormous technical and personal prestige helped to keep tempers down. In trying to solve such problems as shielding a power reactor, insuring thermal stability of a pile, and the like, he showed a first-class engineering talent. He wanted and obtained practical technological results with maximum speed, often using approaches very different from those usually employed by physicists. He knew the value of a speedy answer even if it was approximate or incomplete—and he got it. In some cases his new, unexpected methods gave remarkably accurate results and are now routine in reactor technology.

Physics, however, was always the center of Fermi's interest, and he tried to blend the activities requested by the MED and its war goals with as much pure research as possible. The first pile, as soon as it went critical, became a fascinating physics instrument. Suitably

The War Years

operated, it behaved like a very stable and powerful neutron amplifier, making possible precise measurements of cross sections, which was a central problem for practical applications and also one of considerable theoretical interest. In January 1943, only a few weeks after criticality had been reached, Fermi, Anderson, and Zinn, with their associates, began measuring impurities in commercial graphite and other materials by their influence on the reactivity of the pile. The following months were used in building a second pile at the Argonne Laboratory that was to answer many problems expected in the new technology and that ultimately led to the large production piles built at Hanford, Washington, by the Du Pont Company. The Argonne pile operated at a power of the order of 100 kilowatts, more than 100,000 times the power of the first pile.

Fermi now acquired another regular collaborator in the person of a young woman. This was Leona Woods, who later married the physicist John Marshall and, later still, the chemist W. Libby. She had been one of the workers in Fermi's group from the time of his arrival in Chicago, but now their collaboration became closer. Although she displayed manly vigor and wore simple dress, she was more than pretty and brought a feminine touch to the rather austere atmosphere of the laboratory.

Fermi, Zinn, Marshall, and others did many experiments with the Argonne pile in 1943 and 1944, but the results were classified and not published until after the war. The work forms the starting point of a large and elegant branch of physics in which neutrons are applied to a great variety of problems. It reflects very well the breadth of outlook that was characteristic of the Fermi of pre-neutron days. There is a sort of poetic justice in the fact that Fermi initiated the effective application of neutron physics to solid-state physics. In the theoretical part of this work Fermi systematically used the concept of scattering lengths that he had developed about ten years earlier in a completely different context in explaining a spectroscopic effect discovered by Amaldi and me. The papers are models of simplicity and elegance.

During their investigations in June 1943 the group introduced the device known as a thermal column, a block of pure graphite that acts as a filter, letting through only slow neutrons.[34] With the help of this device one could, by diffraction through a polycrystalline medium, obtain neutrons having a de Broglie wavelength greater than the lattice constant of the medium. This would permit strong

The War Years

beams of subthermal neutrons. In January 1944 the group built a velocity selector,[35] and in the same year they observed refraction and reflection of neutrons.[36] Nevertheless, brilliant and elegant though these applications were, they did not become objects of lifelong fascination for Fermi, as well they might have. He deliberately limited himself to opening the way, leaving to others the detailed exploration of the newly discovered regions.

In addition to the experimental work at Argonne, which probably was the activity Fermi enjoyed most, he had a great number of technical-administrative duties to discharge for MED. He (as well as Wigner and other scientists) found himself in technical disagreement with the DuPont Company over the building of the larger piles. Fermi and H. C. Greenewalt did not see eye to eye on technical-administrative questions. The conflicts of 1943 were serious, although the parties had a very high opinion of each other and ultimately developed a sincere friendship, as is shown by the financial support of Fermi's work by Greenewalt in later years. This also was a critical time because the scientists of Chicago's Metallurgical Laboratory were (for good reasons) worried about the political and social implications of atomic energy for the future. There was a serious ferment of discontent in the laboratory, conflicts with General Groves, and distrust among the younger scientists of the leadership of MED as a whole. No record is available of the part Fermi played in these matters (if any), and he was not as vocal as James Franck, Szilard, and some of the younger scientists, but there are indications[37] that make it very likely that Fermi must have taken a good dose of "physics as soma."

By 1943 the solution of a new and major problem was becoming urgent. If enough U^{235} or Pu^{239} became available, how could one make a bomb? The general idea was simple: one had to assemble rapidly a supercritical mass of fissionable material, and then fission and neutron multiplication in the mass would produce an explosive reaction. The method of assembly, however, and the answers to many other important problems were far from clear. Genuine inventiveness was required before such a weapon could be built. Preliminary studies, carried out at various universities under the auspices of the Metallurgical Laboratory, had produced some of the fundamental nuclear data necessary for planning a weapon.

The War Years

Gregory Breit and Robert Oppenheimer had originally been in charge of this work, and it had proceeded at a modest rate with the help of a small group of physicists, mainly theoreticians, scattered in several laboratories. By the summer of 1942 it was clear that the efforts had to be greatly expanded and that a new laboratory for the specific purpose of assembling a weapon was in order. Safety and secrecy were the most important considerations for choosing the location of such a laboratory, and General Groves, Oppenheimer, and others began to examine possible sites. After an on-the-spot inspection in November of 1942, the choice fell on Los Alamos, New Mexico, the site of a small boarding school for boys. The Los Alamos School was on the rim of a plateau at an elevation of about 7,300 feet. Behind it were high volcanic mountains, pine woods, and meadows; in front, the abrupt edge of the mesa and, below, the plain of the Rio Grande Valley, almost desert except for a fertile strip along the river, dotted by Indian villages. Across the valley the Sangre de Cristo Mountains and Truchas Peak broke the horizon. Apart from the school and a few pleasingly designed buildings of wood and stone, there were no other structures or inhabitants on the mesa, and the dirt road leading to it was primitive. The existing buildings were considered important for a prompt activation of the laboratory. The isolation was an asset; the nearest city, Santa Fe, New Mexico, was about thirty-five miles away to the southeast. The mesa was indented by deep canyons, which in time came to be occupied by special laboratories. The main laboratory and some of the new housing were built on the mesa near the old school immediately after acquisition of the site.

The region, well known to Oppenheimer who used to spend his summers at a ranch in the same general area, was extraordinarily beautiful. This beauty was to have profound influence on many scientists who by inclination and long habit were sensitive to nature and could appreciate the noble country surrounding Los Alamos. The view of the mountains; the ever-varying clouds in the sky; the colorful flowers blooming profusely from early spring to late autumn; the possibility of walking on interesting trails or driving to fishing streams, skiing slopes, mineral beds, or Indian ruins; and many other attractions were to prove one of the greatest assets of Los Alamos and to have a decisive part in sustaining the great effort made by the personnel. Often at the end of a strenuous period of

The War Years

work one felt completely exhausted, but the out-of-doors was always a source of renewed strength.

General Groves, after consultation with important physicists of the MED, appointed Oppenheimer director of the new Los Alamos laboratory.[38] Oppenheimer at the time was professor of theoretical physics at the University of California, Berkeley, and at the California Institute of Technology. He had been among the first to introduce quantum mechanics to America and had founded a flourishing school of theoretical physics which produced many of the leading American theoreticians. He was interested in many subjects besides physics, such as Sanskrit and philosophy; he also was interested in politics and had followed a decided leftist line before the war. His prestige and ascendancy were great among his close entourage, but he sometimes appeared amateurish and snobbish to people more remote from him, who were not under the spell of his personality. He often presented physics in rather abstract terms which contrasted, at least in my mind, with the simple, direct approach to which Fermi had accustomed me. I remember a remark that Fermi made in 1940 at the time of his visit to Berkeley for the Hitchcock lecture. After attending a seminar given by one of Oppenheimer's pupils on Fermi's beta-ray theory, Fermi met me and said: "Emilio, I am getting rusty and old. I cannot follow the highbrow theory developed by Oppenheimer's pupils anymore. I went to their seminar and was depressed by my inability to understand them. Only the last sentence cheered me up; it was: 'And this is Fermi's theory of beta decay.' "

Oppenheimer performed superbly as director of Los Alamos. He was always very well informed on technical details, and he rapidly acquired the necessary administrative experience. He was probably the fastest thinker I have ever met; he had an iron memory and could always speak eloquently on any subject. For all his brilliance and solid merits, however, Oppenheimer had some grave defects, which in part account for the mortal enmities by which he was later unjustly victimized. Very conscious of his intellectual distinction, he was occasionally arrogant and thereby stung scientific colleagues where they were most sensitive; furthermore, he was sometimes devious in his actions. All this bore ugly fruit years later, as I will briefly indicate below. The Medal of Merit, conferred upon him at the end of the war, and the Fermi Award, given to him in 1964 almost in atonement for the terrible ordeal to which he had

The War Years

been subjected in the loyalty hearings of 1954, were deserved public recognitions of his outstanding leadership.

As the very first step, in April 1943, Oppenheimer called a meeting at the Los Alamos school of the scientists who were slated to form the staff of the future laboratory, for the purpose of preparing a program for the impending work. About thirty persons were present. The formal sessions of the meeting were held in the classrooms of the school building; they were followed by spirited discussions, mainly during hikes in the beautiful and savage country surrounding the mesa. The scientists slept in the dormitory of the old school building. Robert Serber wrote a report, now declassified,[39] for new scientific recruits to use as a primer of the problems to be solved. The laboratory was soon to receive a significant portion of the most active and brilliant nuclear physicists of the world. The average age of this group was very low, around thirty-two; a few senior men, already famous, had barely passed forty, and several who would later become famous were in their early twenties. Oppenheimer was thirty-nine years old. Many old friends met again under these unusual circumstances, and the presence of familiar faces (Fermi, E. U. Condon, I. I. Rabi, H. A. Bethe, R. F. Bacher, John Williams, John Manley, Robert Wilson, and others) was a mutual reassurance to all.

Initially, the problems to be solved were presented by different experts, and in the subsequent discussions we made important decisions on the subjects to be studied—on priorities and on methods of approach to engineering and technological questions that were completely new to nuclear physicists. There was not much talk about administrative problems except for the strong objection to putting scientists under military discipline by giving them commissions, a proposal that was never carried out. We expected the enterprise to be shorter, and above all simpler, than it turned out to be, and tried to guess how long it would take to accomplish our mission and how long the war would last. Fermi's help as a consultant had a great influence in defining the problems to be solved and in formulating a program.

After the April meeting, many of those attending, including Fermi, went back to their home bases to prepare the equipment and recruit the personnel needed to activate the Los Alamos Laboratory. Most returned shortly to settle on the mesa, but not Fermi,

The War Years

whose main work was then in Chicago, and who in addition was traveling to Oak Ridge and Hanford. Not until August 1944 could he take part full time in the project at Los Alamos. The building of the facilities proceeded with all possible speed, and by midsummer 1943 the Los Alamos Laboratory began to function. A cyclotron and several linear accelerators brought in from Harvard, the University of Illinois, and the University of Wisconsin formed the first nucleus of experimental equipment. Physicists, chemists, and metallurgists were arriving every day and occupying the new buildings as soon as they became available. Utilities began to function, wooden apartment houses—four families to each house—were built, space allocation problems were solved, a hospital strong in pediatrics and gynecology was built and staffed.

Because of the many children in the community, schools were very important. Scientists' wives, supplemented by a few professionals, supplied the personnel for a school of superior quality for children of all ages. University students accompanying their professors to Los Alamos or sent by the military as members of the Special Engineer Detachment (SED), received a practical scientific education not easily equaled even in the best institutions, supplemented by occasional formal lectures on subjects of interest to the laboratory. Fermi was one of the principal contributors to the educational program for young scientists. Shortly after his arrival at Los Alamos he began to give lectures on many subjects. These were always well attended and a real treat to the personnel of the laboratory. In the fall of 1945, when there was some free time, he taught a regular course in neutron physics; this model course was followed by many of the later nuclear engineering courses.[40]

Administratively, the laboratory was broken down into several divisions: theoretical physics, chemistry and metallurgy, experimental physics, and so on. The divisions contained groups directed by group leaders. The division leaders formed a governing board that met regularly and assisted the director in the establishment of policies and in major decisions. Later Oppenheimer appointed several associate directors to help him in the discharge of his duties. There were many organizational changes during the period from 1943 to 1946, but the general lines remained untouched. Group leaders and division leaders differed considerably in age, experience, and scientific standing, but it was generally understood that no professional slur was meant if it happened that an older, more experienced man was administratively under a younger one. The exigencies of the war justified the arrangements and took away much of the awkwardness of the situation.

Initially the laboratory chiefly measured nuclear properties of the isotopes U^{235} and Pu^{239} as these materials were furnished by other parts of MED. Such data were necessary for planning a bomb and, in particular, for estimating the critical mass. At this time I acquired a special small laboratory for measuring spontaneous fission, the like of which I have never seen before or since. It was in a log cabin that had been occupied by a ranger and was located in a secluded valley a few miles from Los Alamos. It could be reached only by a jeep trail that passed through fields of purple and yellow asters and a canyon whose walls were marked with Indian carvings. On this trail we once found a large rattlesnake. The cabin-laboratory, in a grove shaded by huge broadleaf trees, occupied one of the most picturesque settings one could dream of. Fermi also was very fond of the site and visited us there several times.

During the same time that the nuclear properties of the materials were being determined, other workers were studying methods of assembling the critical mass to produce an explosion. At first the scientists who were charged with this assignment devised rather simple-minded procedures, such as bringing the material together with a gun. Later, after the discovery of the high spontaneous fission rate of Pu^{240} (an unavoidable companion of Pu^{239} as it is produced in a pile), it became imperative to work out more sophisticated methods of assembly such as the implosion method proposed by Seth Neddermeyer and subsequently analyzed and improved by von Neumann and others. Test of the completed bomb was the final wartime assignment of the laboratory.

A very vigorous theoretical division at Los Alamos, under the leadership of Hans Bethe, was composed of Serber, Teller, Weisskopf, D. Flanders, and among the "youngsters," Robert Feynman and Geoffrey Chew. This group, using the nuclear data as they became available, tried to calculate all information that was needed by the laboratory. Lower priority was given to their study of a fusion "superbomb" or "super": it was apparent that such a bomb could not be ready for effective military use for at least several more years, and so it did not claim much effort. This work of the theoretical section involved not only nuclear problems but also

The War Years

very advanced hydrodynamics, to which Sir Geoffrey Taylor, the famous British applied mathematician, and von Neumann brought incomparable expertise. The theoretical nuclear physicists were less familiar with this field, but soon learned much about the subject.

The Los Alamos scientists needed a wide range of information on many phases of the project as a whole, including production schedules; but this need conflicted with the policy of compartmentalization of information, by which one was limited to information of immediate necessity. This policy was one of the pillars of security in the minds of the military authorities, but they recognized that it was incompatible with technical necessities and it was to an extent relaxed for the Los Alamos Laboratory only. Reports from other sites became available in the library, and the transfer to Los Alamos of scientists from other sections of MED brought much oral information to the mesa.

The price we paid for access to information was limitation of personal freedom, something quite new to most of us. From the very beginning, a guarded fence surrounded the laboratories, which were accessible only to authorized personnel, and a second fence enclosed the whole community to control entrance and egress. Furthermore, Los Alamos was very isolated. All trips beyond Santa Fe required specific approval, and mail was censored. These regulations were disturbing, especially to wives of the scientists. The situation is familiar to military families used to garrison life at remote outposts, but it certainly was new to the scientists and their families. One consequence was that many of the residents developed strong feelings about rank and prestige. By hindsight the questions of social prestige rampant at Los Alamos in its early times seem childish, but at the time they appeared very important. All of the new apartment houses were identical, and all were made of wood, with four apartments per building; some of them, however, acquired strange qualities of desirability according to the part of the mesa in which they were located because the address had overtones of social status. A few of the old school buildings, of different construction from the apartments, were in a street nicknamed Bathtub Row because, as the name indicates, the houses had bathtubs instead of the simple showers of the new apartments. Occupancy of one of these houses brought the highest distinction. When the Fermis established themselves at Los Alamos in August 1944, at the suggestion of Laura

The War Years

Fermi they chose an apartment in a part of town that was considered to be of low prestige. They did this to help overcome the silly snobbishness attached to such considerations, and their action had a salutary effect.

The isolation of Los Alamos also pushed families to an active social life: there were many dinner parties; many people for the first time took up poker and square dancing; Teller played the grand piano he had brought to his apartment; somebody else played the trombone; and so on. Fermi participated enthusiastically in these activities.

The visit of Niels Bohr and his son Aage opened a somber view on the state of Europe. Bohr came under an assumed name, but of course many of us recognized him at once. One night, in Oppenheimer's house and in the presence of a security officer, he told a few European scientists of the conditions prevailing in Denmark and of his escape by an open boat.[41] For many of us this was the first eyewitness account of what was really happening in a Nazi-occupied country. Although conditions in Denmark at that time were relatively tolerable and the worse horrors of Nazism were unknown to Bohr, the account left us depressed and worried, and more determined than ever that the bomb should be ready at the earliest date possible.

Fermi was happy in Los Alamos. The work was interesting and had a chance of bringing the war to a victorious end; the company was stimulating; he could again take long hikes in beautiful surroundings and go on skiing expeditions, activities that were very close to his heart. So long as Fermi came to Los Alamos as a visitor, John Baudino, his bodyguard, accompanied him and followed him on our Sunday hikes and on other sporting activities as well. Baudino, however, was not used to long walks, and Laura Fermi tells of an episode that shows the spirit of the situation. One Sunday, apparently because of gas rationing, there was difficulty getting a car to take us to one of the neighboring places from which we wanted to start a hike. Fermi could have commandeered a laboratory car but had scruples about doing this. Baudino the lawyer, however, solved the problem. "You may not be entitled to an official car," he said, "but I have an official duty—to be your bodyguard; and I cannot follow on foot for such a distance. I must, therefore, take a government car."

The War Years

At Los Alamos Fermi took up fishing, perhaps stimulated by the example of many other ardent physicist-fishermen, but he went about it in a peculiar way. He had tackle different from what anyone else used for trout fishing, and he developed theories about the way fish should behave. When these were not substantiated by experiment, he showed an obstinacy that would have been ruinous in science.

Both as a visiting consultant (before he moved to Los Alamos) and later as a resident physicist, Fermi was a sort of oracle to whom any physicist in trouble could appeal and more often than not come away with substantial help. There was no limit to the variety of problems that were brought to him. I remember having listened to Fermi's discussions on hydrodynamics with von Neumann. (These took the strange form of competitions before Fermi's office blackboard as each tried to solve the problem under study first; von Neumann, with his unmatched lightning-fast analytical skill, usually won). I once interrupted one of these sessions because a first-class electronics expert in my group was confronted with a new and very difficult problem in circuitry. The problem was urgent, and in our distress my friend and I dropped in at Fermi's office. In about twenty minutes Fermi devised a circuit that would solve the problem depending however on a tube with a certain special characteristic, but no one knew whether such a tube existed. We consulted a tube manual and found it; the apparatus was promptly built and worked satisfactorily. On another occasion Oppenheimer asked Bruno Rossi and me to prepare a report on the effects to be expected from a nuclear explosion of a certain magnitude. At that time the subject was quite new, and with much labor we tried to estimate or guess at the answers. Fermi happened to be passing by, and we asked him some questions. In no time he clarified the whole subject, and we were able to make a report that later experience proved remarkably accurate, at least as far as physical effects were concerned.

Von Neumann was the other oracle of the laboratory. A distinguished experimental physicist and I had struggled unsuccessfully for an afternoon on a problem that hinged on an integral when, through our open door, we saw von Neumann walking down the corridor toward our office. The integral that baffled us was written on the blackboard. We asked von Neumann, "Can you help us with

The War Years

this integral?" He came to the door, glanced at the blackboard, and dictated the answer. We were absolutely dumbfounded and asked ourselves how he did it. I am sure that similar examples could be quoted by the dozen.

There was friendship and admiration between the two oracles, and their friendship was cemented by a common interest in computers. Fermi had always been adept at numerical calculations, and he saw immediately the possibilities offered by fast electronic computers. He spent many hours learning about them and experimenting with them. The crucial role played by von Neumann in the development of computers is, of course, well known.

In the summer of 1944, when Fermi became a permanent resident of the mesa, Oppenheimer, who wanted to give him an official title, made him an associate director of the laboratory and created a new division, F Division, that he placed under Fermi's direct jurisdiction. The general responsibility of F Division was to investigate problems that did not fit into the work of other divisions. It contained four groups: one under Edward Teller, for theory and especially for study of the hydrogen bomb; one under Egon Bretscher, a British physicist, for experimental help on the hydrogen bomb project; one under L. D. P. King, for building a homogeneous nuclear reactor; and (later) one under H. L. Anderson, for studying miscellaneous problems. The role of general advisor to everybody in the laboratory, which was in many ways the most important function Fermi had, did not satisfy him entirely. He wanted a project he could call his own, and thus he took a direct interest in the building of the homogeneous reactor.

The active part of the reactor was formed by a solution of a salt of uranium 235, in ordinary water, contained in a 1-foot diameter sphere. Thus it was a very small reactor, but it gave a fairly high neutron density. It was first operated at extremely low power, but after a series of experiments it was rebuilt so that it could work at a power of 5 kilowatts. As often as possible Fermi would repair to the Omega site (in a canyon near the main mesa of Los Alamos), where the reactor was located, and spend many hours carrying out experiments and calculations. (One of the members of the Water Boiler group was a very athletic girl, Joan Hinton, a relative of Sir Geoffrey Taylor. Joan became an assistant to Fermi, and with her

The War Years

help he calibrated neutron detectors to be used in the atomic bomb test. At the end of the war she went to mainland China and there married a missionary; she has been in China ever since.)

The Metallurgical Laboratory of Chicago contributed another important member besides Fermi to the directorship of Los Alamos —Samuel K. Allison. Allison's early work had been in the field of X-rays, and many physicists have learned this subject in the famous book written by A. H. Compton and Allison.[42] Allison was brought to the University of Chicago by Compton in 1930, later worked in nuclear physics, and during the war became Compton's right-hand man; when he could be released from the Metallurgical Laboratory, he came to Los Alamos as an associate director of the Laboratory. Allison was a man of great calm, high principle, absolute integrity, and considerable humor, qualities that made him an ideal troubleshooter in difficult personal situations. Because everybody trusted him, he could inject reason into arguments when passion got out of hand, and he was always listened to. He and Fermi soon became close friends. They understood and trusted each other completely, and probably Allison was the man in whom Fermi confided most in the postwar years.

Many famous sayings of Allison illustrate his personality, and his mixture of humor and high purpose has permanently fixed several of his expressions in the folklore of American physics. Reflecting on his experience during the war, he compared a scientist who had been immersed in the university environment to a fish in the ocean. "In the great upheaval of 1940–1946 this particular fish was hauled to the surface in a raging storm, and, groggy from the change of pressure, exhilarated by the oxygen supply, he learned and, equally important, unlearned a great many things."

On September 13, 1944, after much effort, the first production reactor—a water-cooled, natural-uranium pile—with graphite moderator was ready to start operations at Hanford, Washington. The reactor had been built by the Du Pont Company on the basis of a project in which the physicists of Chicago and the company's engineers at Wilmington had cooperated, but not without serious friction. When the reactor was about to be loaded with its fuel, Du Pont called in Fermi to be present for any eventuality. After the success of the first pile at Chicago, Fermi had been present at the start-up of an intermediate reactor at Clinton, Tennessee, which produced gram

The War Years

amounts for nuclear studies. Now, for the start-up of the first large-scale production plant, it was prudent to have "our anchor man" on hand, as Compton wrote.

Why Compton considered Fermi his anchor man is perhaps brought out in a story told by S. K. Allison. Compton, Fermi, and Allison were traveling by train to the Hanford plutonium plant (Compton and Fermi were forbidden to fly during the war because General Groves did not want them to take undue risks, and Allison modestly said that he "was with them for company"). The railroad trip was long and monotonous, and after a while Compton said: "Enrico, when I was in the Andes Mountains on my cosmic ray trips I noticed that at very high altitudes my watch did not keep good time. I thought about this considerably and finally came to an explanation which satisfied me. Let's hear you discourse on the subject." Allison, thinking that this would break the monotony for an appreciable interval, prepared to enjoy the fireworks that would surely follow. Fermi found a piece of paper and a pencil and extracted his slide rule from his pocket. After a few minutes he wrote down mathematical formulas for the entrainment of air by the balance wheel of a watch which would increase the moment of inertia of the balance wheel and hence slow down the watch. He evaluated this effect and came out with a figure that corresponded with Compton's memory of the deficiency of his timepiece in the Andes. Allison said he would never forget the expression of wonder on Compton's face.

The loading at Hanford proceeded regularly over several days, and the reactor started according to plan on September 27, only to lose its reactivity the next day, a few hours after it had reached a substantial power level. The reason for this mysterious failure baffled everybody for a few hours, but then the reactor, having rested, as it were, began to work again. Observing the reactivity as a function of time, Fermi and Wheeler, independently, soon suspected that the effect was due to poisoning of the pile by a fission product, and further study identified the culprit as Xe^{135}. This isotope has a thermal neutron-capture cross section of 3.5×10^6 barns and a half-life of 9.4 hours. Xenon 135 had been found in early studies on fission by C. S. Wu and myself, but it was not known it was such a terrific neutron absorber. Within a few days the physics of the situation was understood, and fortunately, the Du Pont engineers had made provision in their plans for the possibility of adding

The War Years

considerably more fuel than had been calculated by the physicists. They had no justification for this wasteful and expensive precaution, but it saved the day. On this occasion Fermi and Greenewalt worked together for many anxious hours, watching the level of reactivity vary. Fermi was impressed by Greenewalt's force of reasoning in comprehending the phenomena, by which he made up for his rather weak mathematics. Once the Hanford crisis was resolved, Fermi returned to Los Alamos, where another crisis had developed.

The discovery by my group of the high rate of spontaneous fission of Pu^{240} showed that unless a new method of bomb assembly was perfected, pile-produced Pu^{239} would be worthless for use in a bomb. In a pile, Pu^{240} is produced by neutron capture of Pu^{239} and is unavoidable. Pu^{240}, by its high spontaneous fission, produces predetonation and a consequent low yield for a critical mass assembled by the gun method. Thus a major challenge threatened half the activities of MED. Conant's comment after a conference with Oppenheimer, Fermi, and others was: "Alas, and all to no avail."[43]

It was the invention and development of the implosion method of assembly that rescued the project from this critical impasse. A special division, under Captain Parsons, U.S.N., was instituted, and the Neddermeyer group, which had initiated the implosion idea, became part of it. But the work did not proceed as rapidly and smoothly as the circumstances required. Oppenheimer then turned for help to G. B. Kistiakowsky, a Harvard professor of chemistry, and authorized him to prepare a new site on a mesa near Los Alamos for testing explosives. Other ordnance experts also were called in. Tests for the implosion method were carried on by Rossi and Hans Staub, who perfected a technique for measuring the compression achieved. By the end of 1944 the crisis was resolved and the way was clear to use reactor plutonium in a bomb.

No amount of partial tests, however, could replace the definitive test of an entire weapon, a conclusion that had developed over a period of months and was firm at the end of the year. The next question was the procedure for such a test, which was given the code name Trinity.

The first step was the selection of a desert site near Alamogordo for building the necessary facilities. At the same time, at Los Alamos, several physics groups were consolidated into a new organization and began to develop testing techniques. K. T. Bainbridge and

The War Years

J. H. Williams headed the testing organization at Alamogordo and took charge of the complicated logistics. The other physics groups, supported by the powerful organization assembled at Alamogordo, were assigned the many measurements required for the Trinity test, and again Fermi's help was invaluable.

To my knowledge there are no written accounts of Fermi's contributions to the testing problems, nor would it be easy to reconstruct them in detail. This, however, was one of those occasions in which Fermi's dominion over all physics, one of his most startling characteristics, came into its own. The problems involved in the Trinity test ranged from hydrodynamics to nuclear physics, from optics to thermodynamics, from geophysics to nuclear chemistry. Often they were closely interrelated, and to solve one it was necessary to understand all the others. Even though the purpose was grim and terrifying, it was one of the greatest physics experiments of all time. Fermi completely immersed himself in the task. At the time of the test he was one of the very few persons (or perhaps the only one) who understood all the technical ramifications of the activities at Alamogordo.

On May 7, 1945, there was a preliminary test at Alamogordo, with a huge amount of ordinary explosive. This trial run was mainly to calibrate instruments and for practice on an experiment that could not be repeated. The real test would be a one-shot experiment in which failure would be irremediable.

During the May test, when many of the physicists were in the desert, the news came of Germany's surrender and the end of the European war. The reaction of several European physicists was, "Our efforts have come too late." To many of them Hitler was evil incarnate and the prime justification for developing an atomic bomb; now that it could not be used against him, they had second thoughts. These thoughts were not expressed in formal statements, at least not at Los Alamos, but there were many private discussions on the subject.

Nevertheless, the efforts to assemble the atomic weapon were redoubled during late spring and early summer. Material regularly arrived from Oak Ridge and from Hanford and was immediately processed by the chemists and metallurgists. In the meantime the strictly military phase of the project—preparation of the delivery airplane for the bomb's ultimate employment; training of its crews; choice of the objectives—proceeded, under the direction of the mili-

The War Years

tary, at Los Alamos. The military required less help from the scientists on this phase of their assignment, and the scientists were primarily busy with the preparations for the final test at Alamogordo. In July many of the physicists moved to the desert site to prepare for the final tests. The tests consisted in the explosion of an atomic bomb containing Pu^{239}, located at the top of a steel tower, and in the measurement of the energy released under various forms: light, gamma rays, shock wave, and so on.

We established ourselves in barracks, under paramilitary discipline, and Fermi participated in most of the work. He and Anderson also had a special post-explosion project, which was to collect some of the sand and rocks immediately under the tower that supported the bomb and measure the fission products found in the ground. The desert was extremely hot during the day but relatively cool at night, and occasionally there were heavy thunderstorms. In the strange surroundings—scorpions, and gila monsters abounded, and the plants, desert-adapted, looked unfamiliar—the physicists and their helpers ran miles of cables, calibrated innumerable pieces of apparatus, developed routine procedures, and then tested and retested them, always with the thought that there could be no repetition in case of failure, a most unusual condition for an experimental physicist. We worked very hard—best in the early morning and with less vigor as the temperature rose and the light became blinding. In the evening, very tired, we returned to our cots in the barracks. I had brought a French novel to read, a practice I had formed years before as a soldier, and I found refreshment in diverting my thoughts from the present situation to an imaginary world.

Everything was ready for the Trinity test on July 15. On the evening of that day a tremendous thundershower fell on the desert. I had retired, but my attention was attracted by an unbelievable noise whose nature escaped me completely. As the noise persisted, Sam Allison and I went out with a flashlight and, much to our surprise, found hundreds of frogs in the act of making love in a big hole that had filled with water. In the late evening we went to sleep, still wondering if the weather would prevent the test which for many of us was the culmination of years of grueling work and on whose result tremendous issues depended. Oppenheimer, General Groves, and many other authorities occupied preassigned observation points. General Groves has given a description of his and Oppenheimer's activities during the night.[44] I slept and, I assume, so did

The War Years

Fermi, but when General Groves and Oppenheimer decided to proceed with the test, we were called and we went to our preassigned posts. The nearest posts were shelters about six miles from the tower on which the bomb was suspended. Fermi and I were in the next-closest one, at about nine miles, where no special shelter was needed, but as a precaution we lay on the ground facing in the direction opposite to the bomb. Spectators who had no technical participation in the test were much farther removed, about twenty miles from the tower. Sam Allison, from the five-mile shelter, gave the countdown for the electrical impulse that would detonate the bomb, and at 5:30 in the morning the bomb was exploded.

The most striking impression was that of an overwhelmingly bright light. I had seen under similar conditions the explosion of a large amount—100 tons—of normal explosives in the April test, and I was flabbergasted by the new spectacle. We saw the whole sky flash with unbelievable brightness in spite of the very dark glasses we wore. Our eyes were accommodated to darkness, and thus even if the sudden light had been only normal daylight it would have appeared to us much brighter than usual, but we know from measurements that the flash of the bomb was many times brighter than the sun. In a fraction of a second, at our distance, one received enough light to produce a sunburn. I was near Fermi at the time of the explosion, but I do not remember what we said, if anything. I believe that for a moment I thought the explosion might set fire to the atmosphere and thus finish the earth, even though I knew that this was not possible. The margin of safety against such a catastrophic event was sufficient to dispel any rational fear among knowledgeable scientists; however, one can always make errors.

Fermi got up and dropped small pieces of paper on the ground. He had prepared a simple experiment to measure the energy liberated by the explosion: the pieces of paper would fall at his feet in the quiet air, but when the front of the shock wave arrived (some seconds after the flash) the pieces of paper were displaced a few centimeters in the direction of propagation of the shock wave. From the distance of the source and from the displacement of the air due to the shock wave, he could calculate the energy of the explosion. This Fermi had done in advance having prepared himself a table of numbers, so that he could tell immediately the energy liberated from this crude but simple measurement. This anecdote has been told many times, but it is so characteristic of the man that I cannot omit

The War Years

it. It is also typical that his answer closely approximated that of the elaborate official measurements. The latter, however, were available only after several days' study of the records, whereas Fermi had his within seconds. Then, in the first light of dawn, we observed the development of the first mushroom of dust and vapor from a nuclear explosion.

About one hour after the explosion, Fermi donned a protective suit and, carrying a radiation meter, climbed into a specially shielded tank and cautiously proceeded to the site of the explosion to collect materials to be analyzed for fission products. He was much impressed when he found the sand under the detonation point melted to glass. The remainder of the day was spent in collecting the records of the various instruments and in other necessary work. It was late in the evening when we got into the cars that returned us to Los Alamos. Many of us, having had little sleep in the past thirty-six hours, were extremely tired, but fortunately we had soldiers who drove the cars over the mediocre roads. This was one of the very few times in my recollection when Fermi allowed somebody else to drive a car in which he was riding. It was after midnight when we reached Los Alamos and our homes.

The explosion of the bomb had been a success beyond expectation; the energy liberated was clearly near the upper limit, or in excess of our rather dubious predictions. Our satisfaction and pride were great. Although what had happened in Alamogordo was still wrapped in official secrecy, several persons in various parts of the region had seen the tremendous light, and everyone at Los Alamos had an inkling that something very unusual had occurred in the desert. Unofficial photographs of the event had been strictly forbidden, but a young technician in my group, Jack Aeby, had taken color pictures that he now developed in Los Alamos. They were excellent pictures and were ready before the official pictures prepared by the photographic group of the laboratory. Having scooped the official photographers, we went to Oppenheimer and gave him the pictures. Some trouble followed because of the violation of security, but ultimately, I believe, the photographs found their way to President Truman, who was then at Potsdam in conference with Churchill and Stalin.

The explosion of the Alamogordo bomb ended the initial phase of the MED project: the major technical goal had been achieved albeit

The War Years

too late to have a decisive influence on the European front. The feat will stand as a great monument of human endeavor for a long time to come.

The satisfaction and pride of all participants in the work were enhanced by the feeling that its success was bound to bring the war with Japan to an end. At the same time, however, it intensified preoccupations that had beset many participants in the atomic bomb project: What would be the overall consequences of the release of atomic energy and what should one do to direct its immense potentialities to the benefit of mankind?

These were and are tremendous questions that can easily turn into nightmares. Once an individual had decided that the Nazi menace made the attempt to build a bomb necessary, and so long as the pressure of technical work to be accomplished on a tight schedule continued, one could postpone consideration of such questions as less urgent than the immediate task. Also, there had been doubt about the outcome of the work, which after all could have been unsuccessful. Indeed, the margin for success had never been very great, and if some nuclear properties of the heavy elements had been a little different from what they turned out to be, it might have been impossible to build a bomb. Now, however, we had to face these perplexing problems of a nontechnical nature, influenced by our upbringing, traditions, culture, experience, and other widely varying factors. In scientific matters there was a common language and one standard of values; in moral and political problems there were many. Family traditions, early impressions, and schooling often determined unconsciously but powerfully the great diversity of prevailing views. For instance, the Protestant tradition, with a tinge of missionary zeal, firm convictions, and a slight narrow-mindedness, was sufficient to guide and motivate such a man as A. H. Compton, but it had no appeal for an agnostic internationalist with wide experience in several countries, familiarity with diverse cultures, and direct knowledge of European politics, such as Szilard. Nor did it appeal to Fermi. Reconciliation of such differences was difficult because they were based on imponderables, and often on unconscious causes. Furthermore, in science there is a court of last resort, experiment, which is unavailable in human affairs. Authority based on scientific ability was limited to technical fields; it was not recognized in other areas, and justifiably so. It was foolhardy to disagree with Einstein on physics, but quite a different matter to disagree with him on politics. For General Groves the matter was simple: "We had a clear, unmistakable, specific objective. Although at first there was considerable doubt whether we could attain this objective, there was never any doubt about what it was. Consequently the people in responsible positions were able to tailor their every action to its accomplishment."[45] This, of course, is the limited view of a general in the field, as it were; it certainly was not the view of such men as Niels Bohr or James Franck or Leo Szilard. Fermi was not vocal, and even his close friends did not always fully know his viewpoint. His scrupulous regard for security regulations undoubtedly contributed to his reticence. My impression is that he was much less radical than Szilard and more liberal than Compton, Oppenheimer, or Lawrence.

The most thoughtful scientists had from the very beginning of their work on nuclear energy worried a great deal about its implications, but the translation of their preoccupations into words and actions was another matter. It was unclear what one should or could do; political ways of thinking, the historical approach, and the indefiniteness of the problems and of the means to be used are grave obstacles to formulation of policy by a scientist. He is accustomed to thinking along simple if subtle lines, and solving a problem has for him a specific meaning. When he tries to tackle political or social problems, many of the tools on which he is accustomed to rely fail him, and he is at a loss. It was hard to start a dialogue between Bohr and Churchill, as circumstances, alas, proved only too well when the two met in May 1944.[46]

The first persons who became vocal on these questions were members of the Metallurgical Laboratory in Chicago. In the fall of 1943, when the pressure of work relented, they had more time for thinking about such problems as the future of the laboratory and the moral and political issues connected with the release of nuclear energy. The younger scientists, especially, became restless: age and experience had not tempered their sensitivities, and they were worried both by the uncertainty of their own future and by the larger issues. Compton, director of the laboratory, was aware of the problems and of the scientists' agitation. General Groves was informed of this, and Bush appointed a committee, under the chairmanship of Richard C. Tolman, a respected physicist of the California Institute of Technology, to study the future of atomic energy. A similar committee at the Metallurgical Laboratory had been appointed by Compton under the chairmanship of Zay Jeffries, a distinguished metallurgist of the

The War Years

General Electric Co. These committees had both technical and political assignments. Fermi served on the Jeffries committee, which issued a report entitled "Prospectus on Nucleonics."[47] As time went on, the problems became more urgent, more vast, and more difficult. Szilard, Franck, Urey, and many others who were deeply concerned held discussions, proposed lines of action, wrote memoranda, and visited politicians—all this, of course, with proper regard to security.

Fermi was as thoughtful for the future as anyone else, but he recognized the difference between technical decisions, in which he was supremely competent, and political choices, in which he was a normal man well aware of his fallibility. His anchors were his analytical power, confidence in his brain, and his integrity, but he did not like to pass judgment on problems that could not be encompassed by his habitual methods of thought. When confronted by such problems, he preferred not to express himself, especially publicly. This attitude, criticized by some, was rooted in intellectual honesty and in modesty—or, at least, in a perfectly candid appraisal of his limitations. He was, however, not one to shirk his duties; when asked, he would work indefatigably, with total detachment, and consider the issues in the same way that he would consider a scientific problem. This approach often seemed cold and aloof to people who were emotionally involved, especially in comparison with the utterances and behavior of such ebullient personalities as Szilard or Urey. I sometimes thought Fermi believed that when the noise and excitement of the hour had long been forgotten, only physics would last and assert its perennial value. Few men know of Consul Marcellus and the history of Syracuse, but everybody knows of Archimedes.

While Bush and Conant raised broad policy questions with Secretary of War Stimson, President Roosevelt had a meeting with Churchill (September 18, 1944) at which the future of atomic energy, the bomb, and Anglo-American relations in atomic affairs were discussed. A few days after the meeting President Roosevelt called Bush to the White House; but Bush was unaware of the conversations with Churchill, and when the President discussed the S-1 project, he did not reveal his talks with Churchill or their understandings. In the following months Bush and Conant labored on the innumerable, knotty policy points that needed attention;[48] some were of immediate urgency but in retrospect seem of small importance, and others required momentous decisions and were fraught with long-range consequences. Stimson tried to bring these problems to the attention of the president, but the Secretary was overworked and seventy-seven years old, and Roosevelt was tired and had only a few months to live. In reading a summary of this complicated period one can be too easily prone to pass hasty judgment on persons who (to quote Churchill) were "baffled by a problem the like of which, gentle reader, you have not yet been asked to solve."[49]

Death prevented President Roosevelt from acting on Bush's recommendations for a study of the actions demanded by the imminent completion of the Manhattan Project. Vice-president Truman was unaware of the project but was briefed on it immediately after he succeeded to the presidency, and in detail thirteen days later (April 25, 1945) by Secretary of War Stimson and General Groves. A few days later the President appointed a high-level committee composed of Bush, Conant, K. T. Compton, Undersecretary of the Navy R. A. Bard, Assistant Secretary of State W. L. Clayton, James F. Byrnes as the special representative of the president, and Secretary of War Stimson as chairman. The committee had the very broad charge of "recommending action to the executive and legislative branches of our government when secrecy is no longer in full effect."[50] The committee was an "interim committee" because the implementation of its recommendations in many cases required congressional action, which for the time being was impossible because of the secrecy that surrounded the problems. The interim committee appointed a panel of scientists to advise it on technical matters; the panel consisted of Compton, Fermi, Lawrence, and Oppenheimer. This panel had a heavy responsibility indeed, involving, among other things, technical recommendations on the military use of atomic weapons against Japan.

On May 31, 1945, the interim committee met with its scientific panel in Washington, D.C. Stimson opened the meeting by explaining that the committee would make recommendations on temporary wartime controls, public announcements, legislation, and post war organization. Recommendations on military aspects of atomic energy were a responsibility that he and General G. C. Marshall shared. That was why Marshall was present; it was important that he learn at firsthand the views of the scientists. Stimson wanted it understood that neither he nor Marshall considered the project in narrowly military terms. They recognized it as a new relation of man to the universe; they acknowledged that it must be controlled, if possible, to make it an assurance of peace, not a menace to civilization. Stimson hoped the meeting would take a look at the future and consider weapons, nonmilitary developments, research, international competition, and controls.[51] The subsequent extensive and frank discussion lasted the whole day. The scientists made a good impression on Stimson, and Fermi later commented to me on the sharp intelligence and the striking personality of General Marshall. On June 1 the interim committee met with a panel of business leaders, and on the sixth Stimson reported to President Truman on the progress made thus far and discussed specifically the question of international control.[52]

When the scientists returned to their laboratories, they started to work on the assignments they had received in Washington. Also, they had obtained permission to disclose the existence of the interim committee to the members of their laboratories and to seek suggestions and opinions from the scientific community. This was necessary because they were under pressure from their colleagues, and, especially at Chicago, feelings were high and vocal. Of the many suggestions that came to the panel, one of the most important and eloquent was contained in a paper submitted on June 11 by Franck and six of his colleagues, D. J. Hughes, J. J. Nickson, E. Rabinowitch, G. T. Seaborg, J. C. Stearns, and Szilard.[53]

The scientific panel met in Los Alamos during the weekend of June 15–16 in order to prepare its reports to the interim committee. It submitted three papers. In the first paper it recommended that about $1 billion a year be spent by the federal government for fundamental studies on applications of atomic energy. Secrecy should be reduced to the minimum and international cooperation should be sought. The second report recommended that, for the immediate future, the authority of MED be extended to cover work of postwar importance—this work to be done at a cost of $20 million a year.

The most difficult question was treated in the third report, in answer to the question of G. L. Harrison, Stimson's deputy, on the use of an atomic weapon against Japan. The demonstration of an atomic explosion—without a holocaust—that would convince the Japanese to surrender was most desirable. The feasibility of such a plan, however, was doubtful. By June 1945, no weapon had been tested, and the flow of fissionable materials was still very slow. What would happen if the demonstration ended in a fizzle or if, after the demonstration, Japan refused to surrender and there were no materials for rapidly building a weapon? The four men racked their brains trying to determine the technical feasibility of a demonstration on a desert island or in circumstances that would not entail great loss of life. Although I have no direct personal information (the panel activity remained classified until after Fermi's death), I can easily imagine the intense concentrated thought Fermi must have applied to this challenge. The text of the recommendation the scientific panel passed to the interim committee is available only in an accurate paraphrase:[54]

You have asked us to comment on the initial use of the new weapon. This use, in our opinion, should be such as to promote a satisfactory adjustment of our international relations. At the same time, we recognize our obligations to our nation to use the weapons to help save American lives in the Japanese War.

To accomplish these ends we recommend that before the weapons are used, not only Britain, but also Russia, France, and China be advised that we would welcome suggestions as to how we can cooperate in making this development contribute to improved international relations.

The opinions of our scientific colleagues on the initial use of these weapons are not unanimous; they range from the proposal of a purely technical demonstration to that of the military application best designed to induce surrender. Those who advocate a purely technical demonstration would wish to outlaw the use of atomic weapons, and have feared that if we use the weapons now our position in future negotiations will be prejudiced. Others emphasize the opportunity of saving American lives by immediate military use, and believe that such use will improve the international prospects, in that they are more concerned with the prevention of war than with the elimination of this special weapon. We find ourselves closer to these latter views; we can propose no technical demonstration likely to bring an end to the war; we can see no acceptable alternative to direct military use.

With regard to these general aspects of the use of atomic energy, it is clear that we, as scientific men, have no proprietary rights. It is true that we are among the few citizens who have had occasion to give thoughtful consideration to these problems during the past few years. We have, however, no claim to special competence in solving the political, social, and military problems which are presented by the advent of atomic power.

The following weeks were marked by almost desperate activity by many groups and individuals to influence the government's decision on the use of the bomb. Szilard, for one, tried, to talk directly to President Truman but succeeded only in seeing J. F. Byrnes and

The War Years

having a severe misunderstanding with him. Ultimately, the responsibility lay with the President, and although he relied heavily on Stimson's advice, there is no doubt that the decision was his. Whether his decision was right or wrong will be debated for a long time, but after-the-fact judgments are difficult. Certainly the words of Churchill quoted on p. 152 apply most pertinently here. I find it hard to believe that President Truman—or any American President—could have ordered an invasion of Japan, in face of the expected casualties, and banned the use of a decisive weapon that might immediately finish the war. Whether, after Hiroshima, the Nagasaki bomb was dropped too early is another question.

In the summer of 1945 Fermi was faced not only with the momentous decisions and problems mentioned above but also with important personal decisions. With the war ended, he had to think of his own personal demobilization. What would he do and where would he go? He anticipated much future work. He said to me, at the end of the war, that he thought he had accomplished about one third of his life's work. For a time there was talk that Los Alamos might become a university, and if this had happened Fermi might have remained there. The mesa's physical and intellectual environment, its research facilities, and the prospect of teaching its students all attracted him, but the laboratory's future in the undetermined organization of the atomic endeavors became increasingly uncertain. Fermi decided to return to work in a university.

5
PROFESSOR AT CHICAGO

After the war, while science was demobilizing and an uncertainty about the future prevailed in many institutions, the University of Chicago took steps to develop a promising science program; Compton was eager to have the university drop most of its war projects of a technological nature and establish three new institutes: for nuclear physics, for radiobiology, and for the study of metals.[1] An important feature of the new organization was to be a close interaction of the members of the three institutes, patterned on the wartime interdisciplinary collaboration in laboratories such as Los Alamos. His ideas and hopes were shared by other influential members of the faculty. He talked to President Hutchins and the administrators of the university and found them ready to support the plan. By the fall of 1945 the project had made important progress and the university had approached several major scientific figures in the MED to try to attract them to the new institutes. Fermi, Cyril S. Smith (the chief metallurgist at Los Alamos), and Harold Urey accepted three key appointments in the institutes, which were yet to be built. Fermi refused the directorship of the nuclear institute because he did not feel inclined to do administrative work. He successfully urged Samuel Allison to take the job, a move that freed him of administrative duties and at the same time assured a truly superior director for the enterprise. Allison with his exaggerated modesty and dry humor described his job as "the care, mainte-

Professor at Chicago

nance, and appeasement of eggheads." As long as Fermi lived, he and Allison collaborated wholeheartedly, and the character of the relation between the two friends and of Fermi's influence on the institute is well illustrated in the following remarks Allison made on Fermi's death:

> I shall try to express the sentiments of his associates here in the Institute for Nuclear Studies. Actually, the Institute is his Institute, for he was its outstanding source of intellectual stimulation. It was Enrico who attended every seminar and with incredible brilliance critically assayed every new idea or discovery. It was Enrico who arrived first in the morning and left last at night, filling each day with his outpouring of mental and physical energy. It was Enrico's presence and calm judgment, and the enormous respect we had for him, which made it impossible to magnify, or even mention, any small differences among us, such as can arise in any closely associated group. It was at Enrico's personal and urgent request that I took on the chore of directing the Institute in its routine affairs.
>
> It is a completely objective statement, not at all prompted by the emotion of this occasion, to remark that everyone who had more than a trivial acquaintance with Mr. Fermi recognized at once that here was a man who possessed a most extraordinary endowment of the highest human capabilities. We may have seen his physical energy before, or his basic balance, simplicity, and sincerity in life before, or even possibly his mental brilliance, but who in his lifetime has ever seen such qualities combined in one individual?[2]

Fermi accepted the position of professor, effective July 1, 1945, at a salary of $15,000 a year. A strict full-employment clause diverted to the university all income he might derive from consultations, royalties on books, and similar activities. Fermi was scrupulous about the accounting and among his papers are a number of analyses of travel expenses for trips to Los Alamos, Washington, and elsewhere, to determine what was reimbursable and what not. One of his letters contains a pertinent comment on pedantic bureaucracy: "In spite of the fact that I am supposed to have some mathematical skill, I am not able to straighten out in the proper way my expense account for the last trip to Y." When the full-employment policy became optional, he gladly dropped it. His salary was gradually raised, to $20,000 a year in 1951, without the full-employment condition.

The planners of the institutes, desiring to obtain financial support from private sources, needed the help and cooperation of leading businessmen in Chicago. Fermi was prepared to participate in the fund-raising efforts to the extent required by his prominent position in the proposed institute, but his colleagues tried to call on him as little as possible in deference to his clear but unexpressed desires.

On September 1, 1945, when university officials announced the formation of the new institutes, Sam Allison gave a speech in which he strongly attacked exaggerated secrecy and emphasized the necessity of returning to a free exchange of scientific information. If this was hampered by military regulation's, he said, research workers in America would leave the field of atomic energy and devote themselves to studying "the colors of butterflies' wings."[3] Fermi added: "It is not that we will not work for the government, but rather that we cannot work for the government. Unless research is free and outside of control, the United States will lose its superiority in scientific pursuit."[4] These and similar utterances by respected leaders such as Allison, known for his moderation and cool judgment, and Fermi, a rather taciturn and conservative celebrity, made a deep impression on the press and precipitated a swift reaction by General Groves.

Groves tried to discourage further talk by the scientists. He let it be known that a bill dealing with atomic energy was to be introduced in Congress and that public talk might hurt the probability of its passage, but the bill had been prepared almost entirely by the military and its details were not known to the scientists. This fact immediately raised strong suspicions and opposition, and a long, bitter fight developed over the legislation. The scientists—notably the younger scientists at Chicago—promptly formed the Federation of Atomic Scientists (FAS) to further their ideas, which soon spread to Oak Ridge, Los Alamos, and to a lesser extent Berkeley (where Lawrence was unsympathetic). The federation tried first of all to educate the public, including Congress, on the issues involved, and then to assure civilian control of atomic energy and, possibly, obtain international atomic agreements. The opposition, chiefly reflecting the military's thinking, perhaps did not differ greatly in its ultimate goals; however, it was inclined to stress national defense and security and reflected more belief than the scientists had, in the value of "atomic secrets."

Fermi found himself involved in the controversy but, as usual, he looked upon it rather dispassionately and could not get greatly excited about it. Although he certainly was not hostile to them, he did not join the FAS or the sponsors of the *Bulletin of the Atomic Sci-*

Professor at Chicago

entists, the periodical that represented many of the politically inclined scientists and that has done much to enlighten the public on complicated technical matters. His views were expressed in a letter to President Hutchins on September 14, 1945; it is one of the very few public utterances by Fermi on political subjects, and for this reason it is reported here. Hutchins had called a conference on atomic energy, inviting some fifty prominent persons in different walks of life—among them economists and political scientists; Secretary of Commerce Henry Wallace; David Lilienthal, head of the Tennessee Valley Authority and future Atomic Energy Commissioner; and Chester Barnard of the New Jersey Bell Telephone Company. Fermi could not attend, but he sent the following letter:

There is general agreement, I believe, on the following points:

That the new weapon has such destructiveness that in case of a war between two powers both armed with atomic bombs, both belligerents, even the victor, would have their cities destroyed.

The atomic bomb gives an unprecedented advantage to a sudden attacker.

That the balance between defensive and offensive is strongly shifted in favor of the second. Perhaps the only effective defensive measure is a very extensive decentralization of our urban and production centers.

I believe that also the following points are true although the agreement as to them is perhaps less general at least in the non-scientific public:

That secrecy on the industrial aspects of the development would slow up a potential competing nation by only a few years.

That secrecy on the scientific phases of the development not only would be of little effect but soon would hamper the progress of nuclear physics in this country to such an extent as to even make it exceedingly difficult to grasp the importance of new discoveries made elsewhere in the field.

From these points one conclusion emerges. That it is imperative that this country not only should have but should put in operation in a very limited time a policy to face the new dangers. Inaction, hope that things may reach of themselves a satisfactory settlement or engaging in a halfhearted race of armaments would be in my opinion fatal mistakes.

The possibility of an honest international agreement should be explored energetically and hopefully. That such agreement may prove possible is, I know, the most fervent hope of the men who have contributed to the development. In their optimistic moments they express the view that perhaps the new dangers may lead to an understanding between nations much greater than has been thought possible until now.

One of the main reasons why I regret not to be able to attend the conference is that I lose the opportunity to hear the views of people more experienced than I am in international affairs, on the practicability of an international agreement supplemented by effective control measures.

Professor at Chicago

A few remarks as to the peaceful possibilities of atomic energy. There is little doubt that the applications both to industry and to sciences other than physics will develop rapidly. One of the great advantages of an international agreement would be to permit the free growth of such application outside of the shadow of the war use of the new discoveries. Please accept the expression of my regret for not being able to come to the conference.[5]

Secrecy was one of the most important issues. Some of the public had the notion there was a "formula" for the atomic bomb that should be kept secret at all costs. Few persons were so naïve, but on the other hand, only a few persons realized all the consequences of a policy of secrecy. Even among knowledgeable people there were differences of opinion. In August 1945 the government, with the consent of the military establishment, permitted the publication of the Smyth report, an admirably written primer on atomic energy that gave a record of American accomplishments and the basic information necessary for forming an educated public opinion. Its closing paragraph read as follows:

Because of the restrictions of military security there has been no chance for the Congress or the people to debate such questions. They have been seriously considered by all concerned and vigorously debated among the scientists, and the conclusions reached have been passed along to the highest authorities. These questions are not technical questions; they are political and social questions, and the answers given to them may affect all mankind for generations. In thinking about them the men on the project have been thinking as citizens of the United States vitally interested in the welfare of the human race. It has been their duty and that of the responsible high government officials who were informed to look beyond the limits of the present war and its weapons to the ultimate implications of these discoveries. This was a heavy responsibility. In a free country like ours, such questions should be debated by the people and decisions must be made by the people through their representatives. This is one reason for the release of this report. It is a semi-technical report which it is hoped men of science in this country can use to help their fellow citizens in reaching wise decisions. The people of the country must be informed if they are to discharge their responsibility wisely.[6]

The purpose of the report was clearly stated; nevertheless, the conflict with security requirements cannot be glossed over.

Unavoidably, the report gave important and valuable technical information; it described the main lines that had been followed and the success that was obtained. Since these major policy decisions were indeed the most difficult that had to be made during the war,

Professor at Chicago

one cannot deny that the Smyth report had a high military value. Thus its release was inconsistent with a policy of secrecy, but nevertheless it was authorized by the War Department over the signature of General Groves. In the foreword to the Smyth report Groves said:

> The story of the development of the atomic bomb by the combined efforts of many groups in the United States is a fascinating but highly technical account of an enormous enterprise. Obviously military security prevents this story from being told in full at this time. However, there is no reason why the administrative history of the Atomic Bomb Project and the basic scientific knowledge on which the several developments were based should not be available now to the general public. To this end this account by Professor H. D. Smyth is presented.
>
> All pertinent scientific information which can be released to the public at this time without violating the needs of national security is contained in this volume. No requests for additional information should be made to private persons or organizations associated directly or indirectly with the project.

Groves' foreword tries to reconcile the irreconcilable, and the whole report was a difficult feat of tightrope walking. In spite of its release, the obsession with secrecy poisoned international relations, contributed to making the control of atomic energy impossible, and had other deleterious consequences. Secrecy, a card of temporary but great value in the political game, was difficult to play properly. It hampered research, produced embarrassing situations among scientists, and occasionally was manipulated for personal advantage. Fermi's statements on secrecy to President Hutchins were certainly valid.

Legislation on atomic energy was becoming more and more urgent, and President Truman, in a message to Congress on October 3, 1945, formulated broad principles for such legislation. Senator E. C. Johnson of Colorado and Representative Andrew J. May of Kentucky introduced a bill that had been prepared by several experts, chiefly under War Department auspices, giving the military a prominent voice in atomic matters. The sponsors of the bill then held rather hurried hearings, hoping to enact it without delay, but they were completely mistaken in their expectations. Their bill encountered vehement opposition from scientists and some politicians. In contrast Oppenheimer, Fermi, and Lawrence sent a telegram to Secretary of War Robert P. Patterson endorsing the May-Johnson bill.[7]

Professor at Chicago

We would most strongly urge the passage of the legislation now before Congress for the creation of an atomic energy commission. We know from our close association with the actual work in this field that delay will cost us heavily in efficiency, in accomplishment, and in spirit. We believe that with wisdom operations can be carried on within the framework of the proposed legislation safely, effectively, and in the best interests of this Nation. We believe that the broad powers granted the Commission by the legislation are justified by the importance and the perils of the subject. We think it necessary for the American people to understand in full the implications of the new technical situation, but we believe that the proposed legislation will make it possible for their desires and decisions to be responsibly and fully implemented. We assure you that in our opinion the legislation as presented represents the fruits of well-informed and experienced consideration.

J. R. Oppenheimer
Enrico Fermi
E. O. Lawrence

This did not soften the opposition of the other scientists; they could not extend their confidence and respect for the technical opinions of their leaders, to the leaders' political judgment. The hearings, which had to be reopened, produced considerable fireworks, and the fight continued for several months until the May-Johnson bill was replaced by the MacMahon bill. This new bill, emphasizing civilian control, was much more acceptable to the Federation of Atomic Scientists, and in 1946 the federation tried, unsuccessfully at first, to get Fermi and other sponsors of the May-Johnson bill to endorse the MacMahon bill. Fermi gave the reasons for his dissent in writing.[8] He listed ten objections, among them that the bill tended to discourage the development of atomic power; that government would operate installations directly rather than through contractors; that it did not emphasize the role of atomic energy in national safety; that the army did not have sufficient control of purely military applications. Many of these objections were remedied by amendments in preparing the legislation. Before passing the bill Congress debated at length such matters as the composition and powers of the Atomic Energy Commission, research, the production of fissionable materials, security, patent provisions, liaison with the military, and the role of private industry in the applications of atomic energy.

The result is contained in the Atomic Energy Act of 1946, signed by President Truman on August 1, which is still the basic law relating to atomic energy in the United States. The law provides for a civilian Atomic Energy Commission (AEC) composed of five

Professor at Chicago

commissioners appointed by the president and confirmed by the Senate. The commission has two advisory bodies: The General Advisory Committee (GAC), composed of scientists and engineers, which advises on technical questions; and the Military Liaison Committee, composed of representatives of the armed forces, which advises on defense questions. The legislative branch of the government is informed of developments and exerts its influence through the Joint Congressional Committee on Atomic Energy. The first chairman of the AEC was David Lilienthal. The first GAC was composed of Oppenheimer (Chairman); Conant; Lee Du Bridge, president of the California Institute of Technology; Hartley Rowe, chief engineer of the United Fruit Company; Hood Worthington, (soon replaced by O. E. Buckley, president of Bell Laboratories); C. S. Smith; Seaborg; Rabi; and Fermi.

Fermi served on the GAC from January 1, 1947, to August 1, 1950, when his term expired. He asked not to be reappointed because by then the question of the Italian patent had been raised and the suggestion of a conflict of interests irritated him. During the time of his service the GAC met twenty-one times, and Fermi attended eighteen of these meetings. Thus he went to Washington about every other month for from two to four days, the duration of the meetings. However, the time required for preparation was considerably longer, especially for such a conscientious man as Fermi.

I am indebted to Glenn T. Seaborg, present chairman of the AEC, for the following information on the way the GAC operated during the period Fermi served on it. The questions submitted to the committee originated with its chairman, Oppenheimer, with other members of the committee, with the chairman of the AEC, and other members of the commission. They ranged over a very wide spectrum of problems, for example:

1. Policy concerning atomic weapons
2. Policy on the development of thermonuclear explosives
3. Desirable rates of production of uranium 235, plutonium, and uranium 233 for reactor economy
4. Desirable production rates for tracer isotopes
5. Berkeley accelerator proposals
6. Hydrogen bomb secrecy
7. Isotope distribution program
8. International control of atomic weapons
9. AEC contractor relationships

Professor at Chicago

10. Custody of atomic weapons
11. Engineering training
12. Unclassified areas of research

Committee chairman Oppenheimer and Dr. John Manley, the part-time secretary of the GAC, prepared the agenda. Manley, who had been a group leader at Los Alamos, was well acquainted with most members of the committee and had earned a reputation for his ability as a physicist, his tact, and his coolness of judgment. He had been a professor of physics at the University of Illinois before the war, had joined the uranium project very early, and remained with it after the war, except for a period when he served as chairman of the physics department at the University of Washington in Seattle. No minutes of the meetings were kept. After each meeting, usually the next day, a summary letter was prepared by the committee chairman and sent to the chairman of the AEC. Fermi took a leading part in some of the discussions. In the earliest meetings, for instance, he emphasized the importance of keeping Los Alamos a strong laboratory and said that this should have higher priority than the development of power reactors for civilian use. He also emphasized the importance of testing existing nuclear weapons and pursuing the study of a thermonuclear bomb. He also stressed the importance of building a high-flux reactor and pursued this proposal in several meetings, until the plans for building such a reactor were formulated.

Some members of the GAC proposed to establish a new central laboratory or several central laboratories that would be operated by the AEC, but Fermi consistently promoted the view that the present laboratories should be strengthened, even though historical accident, not convenience, had dictated their locations. He strongly advocated balance in the production of fissionable materials—that is, a proper relationship between gaseous diffusion plants and plutonium production reactors. He also advocated, from the very beginning, increased plutonium production capability at Hanford and the design and construction of more technically up-to-date production reactors.

He urged the distribution of radioactive isotopes abroad and the addition of tritium to the list for such distribution—this at a time when opinion was sharply divided on sharing the fruits of nuclear research. He suggested, as early as 1947, that a worldwide network be set up to detect radioactive fission-product debris as evidence of

Professor at Chicago

nuclear-weapon testing by foreign countries. Such a network became an important method for keeping the United States informed of nuclear explosions by the Russians and later by other nations. Fermi also called attention to the importance of delivery problems for nuclear weapons.

On many occasions Fermi emphasized the need for declassification of technical information, criticizing the fetish of security and its many ridiculous aspects. In particular, he stressed the need for enough declassification in the military area to permit sound thinking and planning by responsible military people. He felt there was so much erroneous information about the power of nuclear weapons, their future prospects, their physical dimensions, and similar matters that military planning on a necessarily broad basis was impossible.

One of the major problems was raised on October 29, 1949, under the shadow of the announcement of the explosion of the first Russian atomic bomb, when the GAC was asked for an opinion on how much effort should be put into the development of a hydrogen bomb. At that time the state of the art was primitive and success dubious. A crash program would have diverted so much effort from the manufacture of ordinary atomic weapons as to weaken the military posture of the nation; furthermore the political consequences of an all-out effort did not appeal to the GAC. Their recommendation against the crash problem was unanimous (Seaborg was absent that day). To this recommendation Fermi and Rabi added a minority report which reads:

> The fact that no limits exist to the destructiveness of this weapon makes its very existence and the knowledge of its construction a danger to humanity as a whole. It is necessarily an evil thing considered in any light. For these reasons, we believe it important for the President of the United States to tell the American public and the world that we think it is wrong on fundamental ethical principles to initiate the development of such a weapon.[9]

The AEC decided according to the GAC recommendation, commissioners Strauss and later Gordon Dean dissenting. The AEC decision was attacked by a number of scientists and politicians, while many others defended it. Prominent among the advocates of the crash program were Lawrence and Teller. A bitter dispute, mostly behind closed doors, rent the scientific community and left deep scars.[10] Finally on January 31, 1950, President Truman made his decision and ordered an all-out effort to develop a thermonuclear

Professor at Chicago

weapon. The technical difficulties were staggering. Many of the best minds, including Bethe, Fermi, Teller, and von Neumann, vied in making long, involved calculations to determine whether the then existing schemes to produce a thermonuclear weapon were sound. The results were negative, to the relief of many of the physicists, but in 1951 an invention by Stan Ulam and Teller opened the way to the making of a thermonuclear bomb.[11]

Fermi's work on the advisory committee in the postwar years was his major public service. When his term ended, he was glad. He had worked hard and conscientiously, as always, but without the real pleasure that some people feel in counseling on or deciding paramount questions of national policy. In his attitude toward public service he was at heart a scientist, not a man of affairs.

Let us now leave these matters of politics and administration and return to the *più spirabil aure* ("more breathable air") of science, as Fermi might have said, quoting Manzoni. During the fall of 1945 everyone at Los Alamos discussed at length the impending scientific decisions we would have to make. To Fermi it seemed that, despite —or perhaps because of—its tremendous success, nuclear physics might soon reach a state of maturity and become less interesting to him. He remembered the switch from atomic to nuclear physics about fifteen years earlier and quoted (with his ironic smile) one of the slogans of Mussolini that the Fascists used to paint on Italian buildings: *O rinnovarsi o perire* ("renew ourselves or perish"). Fermi was right, but it took fortitude deliberately to abandon the field in which he was supreme and for which unprecedented tools and facilities were becoming available. Another quotation, from d'Annunzio, covered this point: *Non è mai tardi per andar più oltre; non è mai tardi per tentar l'ignoto* ("It is never too late to go beyond; it is never too late to attempt the unknown"). Nuclear physics could still be pursued, until new facilities were prepared for studying the subnuclear world of elementary particles; but the switch would have to be made, and it was high time to get ready for it by studying all that was known and by building the necessary tools.

Fermi, Allison, and I discussed all this while walking down the Frijoles Canyon toward the Rio Grande, in a strange setting of New Mexico desert cactus and piñons. I do not know whether Fermi was aware of the inventions that would soon make possible accelerators of several hundred MeV. He was interested in machines and fol-

Professor at Chicago

lowed their development, but his attitude was not so much that of a machine builder as that of a user. The physics that could be produced with a machine, not the machine itself, was closest to his heart. Accordingly, he studied and followed with great attention the work on cosmic rays, which at that time were the only important source of high-energy particles.

Fermi and his family left Los Alamos on December 31, 1945, and returned to Chicago. A few months later they bought a large three-story house about forty years old at 5327 University Avenue. The bulk of the furniture came from their prewar home in Italy and included some valuable antique pieces of various periods. Although it might not have pleased the editor of an American magazine on interior decorating, it harmonized with the hospitable personalities and simple tastes of the owners. In the basement Fermi installed a small shop with some power tools. Their daughter Nella attended the college and their son Giulio, the Laboratory School of the University of Chicago.

The Fermis found many friends in Chicago: Allison, Smith, Urey, and Teller had become senior members of the new institutes and lived with their families in the same neighborhood. So did other friends within and without the university. Younger people, such as Herbert and Jean Anderson and John and Leona Marshall among the faculty, and many graduate students, were frequent visitors. Fermi liked the company of the younger generation and participated with glee in games, square dances, and similar activities. The numerous spare bedrooms of their home were often occupied by visitors: old friends, former pupils, and fellow scientists from the United States and Italy made it a point to stop in Chicago on their travels. Lunches at the Quadrangle Club, the University of Chicago's faculty club, offered other occasions for interesting conversations with colleagues and visitors. Fermi was thus continuously informed of what was happening in physics, even without much reading; and conversation became his major source of information. Junior members of the institute (such as Anderson, the Marshalls, Nathan Sugarman, Anthony Turkevich, and others) as well as postdoctoral fellows at the peak of their vigor and full of enthusiasm and zest, helped make Chicago lively. Graduate students who had been recruited from Los Alamos, the metallurgical laboratory, and similar places represented the future. I encouraged Owen Chamberlain and George

Professor at Chicago

Farwell, who had worked in my group at Los Alamos, to finish their education under Fermi at Chicago, and similarly Geoffrey Chew. Marvin Goldberger, and Harold Argo of the Theoretical Division of Los Alamos went to the new institute. The word soon spread, and an extraordinary constellation of students was formed at Chicago, attracted at least in part by Fermi's reputation. The list (though here incomplete) includes H. M. Agnew, R. L. Garwin, David Lazarus, A. H. Morrish, D. E. Nagle, J. R. Reitz, M. N. Rosenbluth, Walter Selove, Jack Steinberger, R. M. Sternheimer, S. Warshaw, Albert Wattenberg, Lincoln Wolfenstein, A. H. Rosenfeld, Jay Orear, R. A. Schluter, H. D. Taft, and G. B. Yodh. From faraway China came C. N. Yang and T. D. Lee. The Institute for Nuclear Studies was primarily a research institution and did not confer degrees, but Fermi also was a professor at the university and had formal teaching duties. He gave regular courses on many subjects, such as thermodynamics, statistical mechanics, nuclear physics, quantum mechanics, and solid-state physics, and he distilled the fruits of his long teaching and research experience into the notes he prepared for these courses. Some of the notes—those, for instance, on quantum mechanics—were continuously improved (The last version, which he completed shortly before his death, was later published and enjoyed wide distribution).[12] Fermi also zealously compiled examination papers for a famous comprehensive examination that was required of graduate students in physics.[13] Every Chicago physics professor formulated questions for this examination, which the students had to answer, in writing, over a period of several days. It was a difficult examination, and Fermi occasionally said with a smile that he was the only member of the faculty who could answer all the questions. (Among his papers I found sets of these examinations that he had answered for his private entertainment; the answer to each problem was usually in one line.) Furthermore, Fermi participated regularly and very actively in two weekly meetings: an informal institute seminar on Thursdays, in which personnel of the institute briefly reported on current research (Fermi frequently volunteered to speak about anything that had crossed his mind—from magnetic fields in the Galaxy to ocean currents or pion scattering); and a more formal theoretical seminar, with preassigned topics and a rapporteur, which met in Gregor Wentzel's office. In the latter the participants went through current developments in theory: quantum electrodynamics, pion theory, and so on.

Professor at Chicago

When Teller was in residence, discussions with him were frequent, and Fermi enjoyed Teller's unusual abundance of original ideas. Fermi often developed these rapidly and far beyond the point reached by Teller—who sometimes missed the pleasure of nurturing his own creatures, although Fermi always gave him full credit for his contributions.

In many ways, it seems to me, Fermi's teaching of his graduate students at Chicago resembled his earlier teaching of our small group at Rome. One of his most proficient disciples, C. N. Yang, describes it as follows:

> It had been my determination, in coming to the U.S. from China in November 1945, to study with Fermi or with Wigner. But I knew that war work had taken them from their universities. I remember that one day, soon after my arrival in New York, I trudged uptown and went up to the eighth floor of Pupin to inquire whether Professor Fermi would be giving courses soon. The secretaries met me with totally blank faces. I then went to Princeton, and found to my deep despair that Wigner would be mostly unavailable to students for the next year. But in Princeton I learned through W. Y. Chang that there were rumors of a new Institute to be established at Chicago and that Fermi would join the Institute. I went to Chicago, registered at the University, but did not feel completely secure until I saw Fermi with my own eyes, when he began his lectures in January 1946.
>
> As is well known, Fermi gave extremely lucid lectures. In a fashion that is characteristic of him, for each topic he always started from the beginning, treated simple examples, and avoided as much as possible "formalisms." (He used to joke that complicated formalism was for the "high priests.") The very simplicity of his reasoning conveyed the impression of effortlessness. But this impression is false: The simplicity was the result of careful preparation and of deliberate weighing of different alternatives of presentation. In the spring of 1949 when Fermi was giving a course on Nuclear Physics (which was later written up by Orear, Rosenfeld and Schluter and published as a book), he had to be away from Chicago for a few days. He asked me to take over for one lecture and gave me a small notebook in which he had carefully prepared each lecture in great detail. He went over the lecture with me before going away, explaining the reasons behind each particular line of presentation.
>
> It was Fermi's habit to give, once or twice a week, informal unprepared lectures to a small group of graduate students. The group gathered in his office and someone, either Fermi himself or one of the students, would propose a specific topic for discussion. Fermi would search through his carefully indexed notebooks to find his notes on the topic and would then present it to us. I still have the notes I took of his evening lectures during October 1946–July 1947. They covered the following topics in the original order: theory of the internal constitution and the

Professor at Chicago

evolution of stars, structure of the white dwarfs, Gamow-Schönberg's idea about supernovae (neutrino cooling due to electron capture by nuclei), Riemannian geometry, general relativity and cosmology, Thomas-Fermi model, the state of matter at very high temperatures and densities, Thomas factor of 2, scattering of neutrons by para and ortho hydrogen, synchrotron radiation, Zeeman effect, "Johnson effect" of noise in circuits, Bose-Einstein condensation, multiple periodic system and Bohr's quantum condition, Born-Infeld theory of elementary particles, brief description of the foundation of statistical mechanics, slowing down of mesons in matter, slowing down of neutrons in matter. The discussions were kept at an elementary level. The emphasis was always on the essential and the practical part of the topic; the approach was almost always intuitive and geometrical, rather than analytic.

The fact that Fermi had kept over the years detailed notes on diverse subjects in physics, ranging from the purely theoretical to the purely experimental, from such simple problems as the best coordinates to use for the three-body problem to such deep subjects as general relativity, was an important lesson to all of us. We learned that *that* was physics. We learned that physics should not be a specialist's subject; physics is to be built from the ground up, brick by brick, layer by layer. We learned that abstractions come *after* detailed foundation work, not before. We also learned in these lectures of Fermi's delight in, rather than aversion to, simple numerical computations with a desk computer.

Besides the formal and informal classes Fermi also devoted almost all of his lunch hours to the graduate students (at least that was the state of affairs before 1950). The conversations in these lunch hours naturally covered a wide range of subjects. We observed Fermi as a somewhat conservative man with a very independent mind. We observed his dislike of pretension of whatever kind. Sometimes he would give general advice to us about our research work. I remember his emphasizing that as a young man one should devote most of one's time attacking simple practical problems rather than deep fundamental ones.[14]

This was Fermi's advice to Yang, and it reflects his own early activities, but the two men made an exception for a paper they wrote in collaboration, "Are Mesons Elementary Particles?" (*FP* 239), which certainly touches very fundamental and deep questions. In it they developed the idea that a meson might be a tightly bound combination of a nucleon and an antinucleon. The idea was probably not entirely new, but the authors tried to develop it quantitatively and to derive from it properties of the nuclear forces. This early incomplete attempt has left a trace in particle physics, and its fundamental ideas echo in later work.

Fermi did not want to limit his teaching to advanced students; he thought that beginners should be brought into contact with science

Professor at Chicago

by active masters rather than by professional pedagogues. He often dreamed of teaching all of physics (and possibly mathematics) to a small group of students from the beginning to the end of their university education. He never had an opportunity to do this; however, he taught the standard elementary physics course to large classes, and with great enthusiasm and success. To my knowledge there are, unfortunately, no records or films of these lectures.

Fermi occasionally spoke of what he would do in his old age. He said that one of his favorite projects was to retire to teach physics in a small Ivy League college and write a book that would contain all the difficult points of physics that are too often glossed over by such phrases as "It is well known." I think he was serious about this, because he started collecting critical questions and even asked me to jot down seemingly elementary questions that I felt I did not really understand. This book would have been a great lesson to physicists, and might possibly have become the all-time best-seller in physics. Unfortunately, however, he did not have the time even to start it.

On his return to Chicago, Fermi regained easy access to the Argonne Laboratory and the unequaled opportunity of using the strong neutron beams produced by its reactors. Because it was only one hour by automobile from the university to the Argonne laboratory, Fermi could work regularly at both places. This was doubly important because the new institute had no buildings or machines as yet. He took full advantage of the facilities at Argonne Laboratory to continue experiments that were a natural sequel of the work he had performed during the war. In a series of investigations with Leona Marshall he explored the phenomena of coherent and incoherent scattering of neutrons. These papers show his full use of the concept of scattering length and a knowledge of solid-state physics that dated back to his student days. The papers are to me striking examples of Fermi's style as an experimentalist. They are not among his most important, but for simplicity and economy of means, both in theory and experiment, and for sheer elegance, they are classics. The subjects he opened by these investigations have developed enormously over the years and now have become separate branches of science.[15]

The immense extension, after the war, of experimental material on nuclei permitted systematic studies of such topics as abundances of isotopes, stability curves, disintegration energies, spins, and magnetic moments. Fermi was interested in these developments and often talked about them with Maria Mayer, who in 1948 was accumulating and systematically examining the experimental material. When she tried to interpret it, she found strong hints of a nuclear shell model, but at first she encountered disagreement between theory and experiment on the "magic numbers" at which the closure of shells occurs. One day, as Fermi was leaving her office, he asked: "Is there any indication of spin orbit coupling?" Mrs. Mayer, who had lived with the data for many months, immediately saw the point and answered without hesitation: "Yes, of course, and that will explain everything." Mrs. Mayer says Fermi was skeptical, but only for a short time. After a week of seeing the evidence interpreted he became convinced and immediately began teaching the shell model in his nuclear physics class.[16] For this work (which was also done independently in Germany) H. D. Jensen and Maria Mayer received the Nobel Prize in physics in 1963.

Ground for the building of the new institutes was broken on July 8, 1947, in the block between Fifty-sixth and Fifty-seventh Streets and Ellis and Ingleside Avenues, on the campus of the university. As soon as the building was completed, Fermi moved to his rooms on the ground floor of the Institute for Nuclear Studies, housed in the south wing of the main building. His office was large, rather dark, and sparsely furnished with a blackboard, a metal desk, and a few bookcases and filing cabinets. In a desk drawer he kept a large supply of spectacles of various strengths that he had bought in variety stores because his eyes had lost much of their power of accommodation. He kept a curve of this power and in 1949 noted that he could read only letters that subtended at least 0.001 radians (the young eye is several times more powerful). His remedy was simple: when eye accommodation was required, he picked from his drawer the glasses best suited for the distance at which he wanted to read or see. He also tried to remedy a much less apparent failing of his memory by an "artificial memory," as he called it. He had accumulated this "memory" over many years, choosing significant reprints, numerical data culled from the literature, and many of his own calculations written in clear handwriting on loose sheets of paper. The documents were numbered and classified with elaborate cross-references and finally filed in cabinets and indexed in a notebook. If he needed, for instance, such an item as the electric and magnetic fields corresponding to electromagnetic waves in a cylindrical cavity, or the equation of the state of matter under extreme temperature and pressure, or numerical data on the solar system, or what Feynman had said at the Pocono Conference, he would look at his booklet and in a few seconds could produce the desired information from the cabinets, often in the form of his own personal calculation on the subject. Adjacent to his office he had a large laboratory that contained tools, a chemical hood, benches, instruments, a drafting table, and so on.

In the meantime the big accelerators of the postwar era, using the phase stability principle, had started operating. Berkeley's 184-inch cyclotron first functioned in November 1946, and in 1948 it produced the first artificial mesons by bombarding severel elements with alpha particles. The Berkeley synchrotron produced mesons by photelectric effect in 1949. Other smaller machines, producing 200-MeV protons, were in operation at Rochester University and at Columbia. The Berkeley results on pions were very impressive; the discovery of the π°, precise determination of the pion masses, interaction with hydrogen and deuterium, and mean-life measurements were major results of the initial operation.

The physicists in Chicago were anxious to have their own machine, and they planned a 170-inch, 450-MeV synchrocyclotron, to be housed in an accelerator building attached to the institutes, but the cyclotron was not to produce its first beam until the spring of 1951, and while waiting for the event Fermi could not do much experimental work; instead he did theoretical work on particle physics and related subjects.

Immediately after the war the Rockefeller Foundation and the National Academy of Sciences sponsored several private conferences on problems of immediate interest to theoretical physics. (These were probably modeled on the historic Solvay conferences of the prewar era.) Oppenheimer (who in April 1947 had accepted the directorship of the Institute for Advanced Study at Princeton) was one of the active organizers. Fermi attended the second conference (held April 2–4, 1947, at Shelter Island) on the foundations of quantum mechanics.

It was at this conference that Willis Lamb gave the first account of preliminary measurements of the level shift in hydrogen (now known as the Lamb shift) and Rabi mentioned Kusch's experiments that indicated an anomaly in the magnetic moment of the electron.

Professor at Chicago

Immediately after the conference Bethe (reportedly while returning by train to Ithaca) made the first calculations that explained the Lamb shift by mass renormalization. Fermi followed the developments carefully but did not participate in them. The third conference, held March 30–April 1, 1948, at the Pocono Manor Inn was especially memorable for Schwinger's presentation of his ideas on electrodynamics. His talk, which lasted many hours, required an extraordinary effort of sustained attention, and one by one the great physicists succumbed to fatigue (or so Fermi thought from their looks); only Bethe and Fermi could follow Schwinger to the end. Fermi chuckled while telling me of this endurance test. Back in Chicago, in an effort to master the new techniques, he wrote extensive notes on various forms of quantum electrodynamics and a number of applications that he later expanded to encompass Feynman's methods.

The year 1947 brought a new and important result in high-energy physics: Conversi, Pancini, and Piccioni—who during the war had worked in a cellar in Italy while hiding from the Germans —published their observations of the different behavior of positive and negative mesotrons (as they were then called) coming to rest in iron or graphite. They observed that, in carbon, both the positive and negative mesotrons decayed and emitted (respectively) electrons or positrons, while in iron the negative mesotrons were captured before decaying. The experimental result was communicated to Fermi before its publication. According to the views then prevailing, the negative mesotrons should have been captured by the carbon nuclei long before they disintegrated, and the experiment pointed to a fundamental discrepancy between the cosmic-ray mesotrons they had observed and the expected behavior of the Yukawa particles connected with nuclear forces. Fermi and Teller, together, and Weisskopf, independently, were quick to grasp the far-reaching consequences of the experimental result. In February 1947 the three together wrote a short note on the subject,[17] and later Fermi and Teller analyzed minutely the slowing-down process of mesotrons in solids to make sure there were no flaws in their arguments.[18] These difficulties were mentioned at the Shelter Island conference, where Marshak (as had been done previously and independently by Sakata and Inoue) hypothesized the existence of two kinds of mesons. The paradox was soon resolved by the discovery of the pi-meson and by the recognition that the μ-mesons, or muons, were its decay product.

Professor at Chicago

The origin of cosmic rays had always interested Fermi. He had talked about the problem to Amaldi as early as 1939 and had expressed some ideas on the acceleration of protons as a tendency to equipartition of energy. It was obvious, however, that some links were missing. With what were the protons colliding? During a visit to Chicago in 1948, Hannes Alfvén informed Fermi of the possible existence of extended magnetic fields in our galaxy. This was the missing element that was necessary to bring to fruition Fermi's previous ideas. Fermi made the hypothesis that protons accelerate by colliding with the extended regions of space which contain the magnetic field and act almost as material objects, and that the energies observed are reached in an attempt to establish equipartition of energy between one proton and the energy contained in a vast region of space. Although this is not the final, complete theory on the acceleration of cosmic rays, the ideas Fermi developed are valid and have influenced all subsequent development of this subject.[19]

With the return to normal academic life, Fermi resumed the habit of spending his summer vacations away from his university. Many institutions were eager to have him visit them, and at almost all of them he had made personal friends, so that the only problem was choosing among the various invitations. He reserved a few weeks every year for the Los Alamos Laboratory; his duties as a member of the GAC, although they did not require such visits, rendered them advisable. Furthermore, his many friends at Los Alamos, the attractive surroundings, and one of the best computers of the time also drew him back to New Mexico. He also went to the University of Washington in Seattle (in 1947), to the University of California at Berkeley (in 1948), and to Brookhaven National Laboratory (in 1952). During these visits the many formal and informal talks, picnics, hiking expeditions, and social gatherings provided inspiration for both visitor and host and often, indirectly, bore important scientific fruits.

In the summer of 1953, to my surprise, Fermi spent a week in Aspen, Colorado, as a participant in an adult education program for about twenty young businessmen, directed by O. M. Wilson of the Fund for the Advancement of Education of the Ford Foundation. Another physicist, Lee Du Bridge, also attended these seminars, and the discussions were lively. They centered around selected readings, among them documents of American history, and excerpts from classics of literature, philosophy, and the social sciences. Fermi talked a great deal and later commented to me about his surprise at being considered an eminent philosopher and being listened to respectfully by the rest of the group: "It seems that I am a great philosopher, without knowing it," he said, shaking his head. But this was not his first experience of this type: he and his wife belonged to a book group in Chicago that discussed such authors as Ortega y Gasset, Maritain, Tillich, Whitehead, and Bertrand Russell and such topics as western and oriental religions, ethical questions, ancient anthropology, and the development of man. Fermi expressed strong opinions and often tended to dominate the discussion; sometimes he began by saying, "I have not read the book, but—" (he was not the only one to use this opening, however). Occasionally, as he told me, he would win a point by the old Italian method of shouting louder than his opponent. These experiences and the steady reading he had been doing since coming to America (a short session every evening before going to sleep) extended and deepened his cultural interests beyond what they had been in his Roman days. Occasionally, he hinted to me, he meditated on epistemological aspects of quantum theory, and he once remarked to his wife that "with science one can explain everything except oneself." The incidents suggest that, like other eminent quantum theorists, he had to wrestle with the problem of self-consciousness.[20]

In 1949 Fermi returned to Europe, the first time since his departure more than ten years earlier. The University of Basel had called a conference on high-energy physics, to be followed immediately by a conference in Como on cosmic rays. Fermi attended both, and at Como he reentered Italy for the first time since 1938. Many of his old Italian friends were there: Amaldi, Bernardini, Pontecorvo, Occhialini, Rossi, Wataghin, and I; and many young Italian physicists, who met him for the first time. At Basel Fermi lectured on the origin of cosmic rays, and swam a mile in the Rhine; at Como he played tennis with Pontecorvo, who was almost a champion. It was amusing to watch Fermi compete strenuously with somebody clearly superior to him—and lose.

After the Como conference he gave a series of nine lectures, which were sponsored by the Accademia dei Lincei under the Donegani Foundation—six in Rome and three in Milan.[21] The new generation of physicists, grown in Italy after his departure, saw and heard the almost legendary Fermi, who brought them in a simple

Professor at Chicago

but original and profound form some of the new developments in topics which were at the forefront of interest: elementary particles, theories on the origin of the elements, neutron-electron interaction, nuclear shells, quantum electrodynamics, the neutron, neutron optics, and Dirac's monopole. The lectures were heard by members of a postwar generation of physicists several of whom were then just beginners in research but were later to become well-known scientists. The venerable Professor Castelnuovo, president of the Accademia dei Lincei and a member of it before Fermi was born, presided at some of the lectures. Castelnuovo, who had known and befriended Fermi in his student days, had been dismissed from his mathematics chair and from the academy in 1938; and under Nazi occupation his life had been in danger. The reconstituted Accademia dei Lincei—it had been disbanded by the Fascists in 1939—elected him as its president, and his presence at Fermi's lectures was an impressive symbol of Italy's rebirth.

After his return to Chicago, Fermi continued his systematic preparation for experiments in high-energy physics. He knew the limitations of the available theoretical knowledge but made a determined effort to calculate as many crosssections, half-lives, etc., as possible, separating the known factors (such as those deriving from phase space) from unknown matrix elements. Thus he prepared, on the basis of what was known, the forms—as it were—that would be filled in by the imminent experiments. There are traces of this work in the Silliman lectures on elementary particles that he gave at Yale University in the spring of 1950.[22] (Fermi once compared this work to what one could have done in the age of Lorentz, before the discovery of the quantum, to explain the atomic optical phenomena.) In the same vein are some papers in which he developed a statistical theory of high-energy events.[23] He assumed that in a collision the strong interactions establish statistical equilibrium in a certain volume, and he derived many consequences from this hypothesis. The work has been useful as a qualitative guideline in the preliminary planning of experiments.

The convocation, in 1950, at Rochester, N.Y. of the first conference devoted entirely to high energy phenomena, indicates the growing interest of physicists in this subject. A relatively small number of scientists attended, and the meeting was mainly devoted to cosmic rays. The organizer, Robert Marshak, managed to de-

Professor at Chicago

velop a very useful and agreeable working conference, so successful that the conference became an annual event attended, by invitation, by the foremost workers in the field. Fermi attended them regularly and frequently intervened in the lively discussions following the presentation of the papers. The informal atmosphere and the youthful, enthusiastic attendance spurred him to give his best performance.

In the spring of 1951, finally, a beam emerged from the Chicago cyclotron, whose construction had been directed by Anderson and John Marshall. Fermi had followed the progress of the construction closely, even building with his own hands a movable target that proved very useful, and calculating orbits for the emerging beam with an analogue computer of his own invention and construction. To inaugurate the cyclotron the University of Chicago called an International Conference on Nuclear Physics and the Physics of Fundamental Particles from September 17 to 22, 1951. The conference fell just before Fermi's fiftieth birthday, and several friends had wanted to celebrate it, but there was only a breakfast with intimate friends and colleagues. At the meeting the Chicago physicists reported the first experimental results from the new cyclotron.

At this point of his life, Fermi resumed his role of experimental physicist, not to abandon it until his death. The problem he attacked —in collaboration with H. L. Anderson, Arne Lundby, R. L. Martin, D. E. Nagle, G. B. Yodh, and several staff members and students—was the pion-nucleon interaction, which is the fundamental step in the Yukawa picture of strong interactions. This was the main subject of Fermi's investigations for the three remaining years of his life. The study required a strenuous effort planned on a large scale, and Fermi pursued it with all his usual vigor.

In our frequent reunions of those years—in Chicago, during summer vacations, at Rochester, and elsewhere—I had the impression that Fermi was becoming more and more time-conscious. I noted a concentration on work of immediate interest and an unusual detachment from physics that was removed from his main topic of study, a specialization that was uncharacteristic of him. I occasionally had the impression that he was constantly aware of the limited time given humans and was determined to accomplish the maximum within his powers. I never heard him say that he did not want to waste time, but his acts, his expression, and all his behavior indicated this in such a way as to inspire reserve or caution in everyone who would confer with him.

One of the main results of his work during those three years was the experimental verification of the conservation of isotopic spin and the recognition of the first pion-nucleon resonance, the one having spin 3/2 and isotopic spin 3/2. Keith Bruckner had proposed these ideas and written a paper on them. A preprint came to Fermi, who—according to Anderson—read it while doing experiments at the cyclotron. Fermi immediately grasped the importance of the idea. He left the experimental room, and after about twenty minutes spent doing calculations in his office, he returned with the announcement that the cross sections for elastic pi plus-proton scattering, pi minus-proton charge exchange scattering, and pi minus-proton elastic scattering would be in the ratio 9:2:1, a prediction that was soon confirmed by his experiments. This was the first discovery of a pion-nucleon resonance;[24] many more followed, at higher energies, but Fermi did not live to see them.

The new computers then coming into use were admirably suited for analysis of the pion-nucleon scattering experiments because they allowed complete utilization of the experimental data, a superhuman task without such powerful computational aids. Fermi managed to spend some time at Los Alamos, where an electronic computer was available, and the method, which he developed in collaboration with Nicholas Metropolis, is now the standard tool for phase shift analysis.

In the last two years of his life Fermi again participated in public affairs. I do not know by what arguments he was persuaded to stand for election as vice-president of the American Physical Society, but perhaps the beginning of the McCarthy era, which posed a menace to science, suggested Fermi's draft to the nominating committee, and he felt he had a duty to serve.[25] The presidency of the society automatically followed in 1953, just when the office (normally rather aloof from controversies and to a large extent ceremonial) was suddenly forced into a polemical position.

The Secretary of Commerce in the new Eisenhower cabinet, although in charge of the National Bureau of Standards, had little understanding of its function. The bureau had reported to the Post Office Department its finding on a battery additive (AD-X2) whose manufacturers had made unsubstantiated claims for it, and the product had been banned from the mails as fraudulent. The Secre-

Professor at Chicago

tary of Commerce, under political pressure, overruled the bureau and said that the "judgment of the marketplace" was more important than scientific tests. The unwise statement and the suspension of Allen V. Astin as director of the Bureau of Standards created a furor. The National Academy of Sciences and the American Physical Society intervened in the controversy, and the Secretary of Commerce countermanded Astin's suspension. The following statement of the Council of the American Physical Society sets forth the position of that body and of its president, Fermi:

> The Council's apprehension has been partly relieved by the temporary reinstatement of Dr. Astin, and by the accompanying declaration of Secretary Weeks that his previous action did not imply any reflection on the scientific integrity of either Dr. Astin or the Bureau of Standards. However, more is needed to undo the harm that has been done; and a fundamental principle should be made very clear.
>
> It is the duty of a scientist to investigate scientific and technical problems by openly-stated objective methods without shading his conclusions under political or other pressures. On this principle the progress of science depends. We never doubted that the work of the Bureau of Standards has been conducted in this spirit.
>
> The Council urges that an authoritative statement be made that this principle forms the rule of ethics for scientists in Government service and that no scientists will be penalized for adhering to them. We believe that such a statement would do much to relieve the uneasiness caused by the Astin incident.[26]

Later in August 1953 the Secretary of Commerce, realizing his blunder, permanently reinstated Astin, but at the same time the Post Office ban against AD-X2 was lifted.

The Astin affair had barely subsided when a graver one arose, one in which Fermi again played a part and whose consequences troubled him through the last weeks of his life. In November 1953—at a time when Congress, the administration, and the American public seemed obsessed with security—L. W. Borden, formerly executive director of the staff of the Joint Committee on Atomic Energy, wrote a letter to the director of the FBI, J. Edgar Hoover, in which he stated (among other things) that in his opinion "more probably than not," J. Robert Oppenheimer had been a spy for the Russians in the period 1939–42. These charges were never pressed, but Hoover referred the letter to President Eisenhower, who conferred with Admiral Lewis Strauss, the newly appointed chairman of the Atomic Energy Commission; and Strauss announced that the AEC

Professor at Chicago

would review Oppenheimer's clearance in accordance with new regulations that had come into effect. Pending the outcome of this review, a "blank wall" denied Oppenheimer access to classified information. All this caused extreme surprise in the scientific community, and a bitter pro-and-con division soon developed. Oppenheimer had been very much in the public eye; his merit had been recognized by the government and the public at large; and now, suddenly, after a decade of service in high places, he was declared untrustworthy. There were many causes for this stunning, sudden development: the climate of the times made people morbidly sensitive to issues of security; Russia's recent progress in atomic weapons also stimulated fear; but, unfortunately, private grudges and personal jealousies also entered the picture. One was reminded of the Dreyfus affair in France. Oppenheimer requested a review of the order that deprived him of his security clearance, and this produced hearings before a Personnel Security Board of the AEC composed of Gordon Gray, Ward Evans, and Thomas Morgan, from April 12 to May 6, 1954. Approximately forty witnesses, several of them prominent scientists, testified under oath. The testimony was to be kept secret, but this decision was reversed and the 992-page transcript was published within a few months.[27] The book still makes fascinating, though often disturbing, reading. One can see distinguished scientists carried away by passion beyond what justice and fairness would have required. One can see persons in relatively high places making foolish utterances; while some witnesses showed great force of character, honesty, and intelligence, others blatantly showed human weakness.

Fermi's testimony, given on April 20, 1954, was relatively brief, and he complained that it had been cut short (perhaps to accommodate the next witness). Had Fermi had more time on the stand, and had the lawyers questioned him in this direction, he would have emphasized the services rendered by Oppenheimer. The testimony is very much to the point, unambiguous, and factual as one would have expected of Fermi. It shows his clear mind, upright character, and judicial thinking. On its basis Oppenheimer's clearance would never have been revoked. It was his opinion that Oppenheimer had rendered outstanding service during the war, that after the war his advice had been given after thorough study and in good faith. If it had not been taken, or if it was thought to be wrong, these facts offered no grounds for impugning his loyalty. Oppenheimer's past had been investigated and considered acceptable before and during the time he was director at Los Alamos and should not be questioned again in the absence of new facts.

The three-man security board decided, two to one, that Oppenheimer was a security risk, according to the definition of the Eisenhower directive, because he had shown bad judgment in opposing the crash program for making a thermonuclear bomb in 1949 and because he had displayed a lack of enthusiasm for the project. In his dissenting opinion, W. V. Evans, a chemist and member of the board, said that an advisor had to give advice according to his best judgment, and he saw no evidence that Oppenheimer had in any way hindered the execution of the presidential order for developing the bomb after it had been given. This dissenting opinion was close to Fermi's thinking. The decision of the security board was upheld by a four-to-one vote of the commissioners—commissioner Smyth dissenting—but on different grounds: the commissioners decided not on Oppenheimer's recommendation concerning the hydrogen bomb, but on incidents persuading them that he had exhibited weakness of character and had entered into association with unreliable persons.

Fermi was deeply saddened by all this. He deprecated the passions aroused in the dispute, which impeded a fair judgment of the issues, and the deleterious divisive effect of these passions on the scientific community. One of the last times I saw him, at the hospital, when he knew he had very little time to live, he said that he wanted to set straight a friend whose testimony he thought had been unethical. He smiled with slight irony and said, "What nobler thing for a dying man to do than to try to save a soul?" A reflection of his feelings on the entire matter is found in a press interview that he gave at the hospital, but which was published in a much edited form by *Time* magazine.

In 1953 Fermi had performed his last experiment on pion-nucleon scattering, and in the summer of that year he analyzed the data at Los Alamos. In 1953 and 1954 he wrote a few theoretical papers on the origin of cosmic rays, on multiple production of pions, and on an ingenious application of computers to a theoretical experiment concerning solutions of a nonlinear vibration problem.[28]

Professor at Chicago

In the meantime, experiments on proton polarization in high-energy scattering were starting at Rochester, Berkeley, and Chicago.

In February of 1954 I visited Fermi in Chicago for a few days and described the recent results of the Berkeley experiment on proton polarization. We had discussed this problem in November of 1953, when the Rochester group had obtained polarization, but a Chicago group had been unable to observe it. I told the latter that on the basis of our experience the reason for their negative result seemed to be an unhappy choice of the scattering angle. By February, at Berkeley, we had much better experimental results with quantitative measurements, and Fermi was curious to see whether the spin-orbit coupling, which plays such an important role in the shell model, could also account for the polarization in high-energy scattering. During my visit he calculated the effect on his office blackboard, working on the problem from ten in the morning to about noon. He first made a false start in using the Born approximation, which gives a null result, but immediately corrected it and proceeded rapidly while I took notes, which with minor changes he used as a draft for the paper "Polarization of High Energy Protons Scattered by Nuclei" (*FP* 267). Fermi liked the simplicity of the argument and the results, and lectured on them that summer in Italy. This was the last time I saw him solving a problem in the old style, so familiar to me from the Roman period. Probably his last published original work, it preceded his final illness by only a few months.

During the summer of 1954 Fermi again went to Europe. He had prepared a beautiful course on pions and nucleons,[29] which he delivered at the Villa Monastero (in Varenna on Lake Como) at the summer school of the Italian Physical Society, which is now named after him. He also visited the French summer school at Les Houches near Chamonix, and lectured there. But his health was failing; an insidious illness had attacked him and could not be diagnosed in spite of repeated examinations. With tremendous willpower he tried to pursue his ordinary life, including mountain hikes and sports, but when he returned to Chicago he went to Billings Hospital for a thorough investigation. By then there was evidence of a serious stomach disease and an exploratory operation showed a malignant tumor that had metastasized, a hopeless condition.[30]

I had just returned from a trip to South America when I received

Professor at Chicago

a telephone call from Sam Allison, who in a broken, almost unintelligible voice told me of the operation that had been performed that morning and its result. I did not know that Fermi had not been well, but the tone of Allison's voice instantly revealed the truth. I went to Chicago as soon as possible. Fermi was resting in the hospital, with his wife in attendance, and was being fed artificially. In typical fashion he was measuring the flux of the nutrient by counting drops and timing them with a stopwatch. It seemed as if he were performing one of his usual physics experiments on an extraneous object. He was fully aware of the situation and discussed it with Socratic serenity. He commented on his conditions, on family questions, on the Oppenheimer affair, and on the future of science and mankind. He said also that if the disease would leave him the time for it he would write his course on nuclear physics as a last service to science; and in fact when he went home some time later, he tried. An unfinished page, a table of contents for this course, is his last writing.[31] He preserved to the last an almost superhuman courage, strength of character, and clarity of thought.

Fermi died on November 29, 1954, just two months after his fifty-third birthday, and was buried in Chicago.

It is difficult to assess the place of Fermi among the scientists of modern times. The perspective is too short, and such a judgment is even more difficult to a friend and pupil. Fermi's greatest achievements, chronologically, are: the discovery of Fermi's statistics, the beta ray theory, and the experimental neutron work beginning in Rome and culminating in the chain reaction.

The Fermi statistics (independently discovered by Dirac) was the key to the modern theory of metals and the statistical models of the atom and nucleus and is basic in many parts of physics. Pauli's principle, however, is the fundamental discovery in this field.

The beta ray theory increases in importance with the passing of time. It was influential in introducing field theory to elementary particles; it proved to be far sighted in the choice of the vector interaction; and it has been the inspiration for much later work. It might well be Fermi's greatest contribution to theory.

The neutron work contains the fundamental discovery of slow neutrons and the crowning accomplishment of the chain reaction, a milestone in the history of mankind.

Professor at Chicago

I have here omitted other scientific accomplishments of Fermi which would be sufficient to make lesser physicists famous.

Fermi's influence on physics in Italy can hardly be overestimated. He was the founder of a movement which rapidly brought Italy from a backward place to an important position in the world of physics. His influence in the United States, though great, was less unique.

Fermi should not be compared with Maxwell or Einstein, who were in a class of their own, and comparison with more recent scientists is difficult from our point of vantage.

In any case Fermi gave science his utmost, and with him disappeared the last individual of our times to reach the highest summits in both theory and experiment and to dominate all of physics.

APPENDIXES

Appendix 1
LETTERS TO ENRICO PERSICO

The following letters from Fermi to Enrico Persico are reproduced with the kind permission of Professor Persico. The translation is by E. H. Segré. After 1926 the two friends corresponded less regularly, and later letters are not available.

Rome, September 7, 1917

Dear Enrico,

I terminated the bathing season at Ladispoli about a week ago and I gladly accept your kind invitation. I will be in Frascati on the first train, and will leave Rome at 6:30 A.M. on Monday the 10th. I do not come on Saturday because I probably would not have the time to notify you of my arrival.

I go every morning to the Vittorio Emanuele Library. A few days ago I went to visit Professor Eredia[1] to calibrate the barometer,[2] but I have not done the calibration because, on the advice of the professor, I will take seven or eight readings which I will then compare with the observed pressures in order to obtain a more accurate value of the mean. I urge you to study the history of Napoleon carefully.[3]

With many thanks, a thousand greetings to you and best regards to the family—also from my mother.

Enrico Fermi

Rome, August 18, 1918

Dear Enrico:

I told you that I was to go to Piacenza at this time, but now the departure has been postponed for about a fortnight. I am sorry that just as you arrive I will be leaving. The reading of Chwolson[4] proceeds quickly and I anticipate that in three or four days it will be finished. It is a study I am very glad to have made because it has deepened the notions of physics I already had and has taught me many things of which I didn't have the slightest idea. With these foundations I hope that I will be able to compete for Pisa with some

Letters to Enrico Persico

probability of success. If I should then accept, *"ghe pensarum."*[5]
Best greetings to your parents and a friendly handshake from
Enrico Fermi

Pisa, December 9, 1918

Dearest Friend:

I am slowly adapting myself to my new surroundings. I must acknowledge that during the first days of the new life I was slightly despondent. However, everything has now passed and I have completely regained my self-control. Let me receive news of you often—this will always be highly welcome. My address is: Enrico Fermi, Scuola Normale Superiore, Pisa. How are your studies proceeding? And the horizontal component of the earth's magnetic field?[6] I end my postcard because I must go to the calculus lecture.

Best regards to your mother and father and a cordial handshake from your friend,

Enrico

Pisa, February 12, 1919

Dearest Friend:

I have finally succeeded in unraveling that business of the circular circuit. As I suspected, it is a resonance phenomenon which a conductor of that shape presents for suitable wavelengths, depending on the geometry of the circuit. The phenomenon is of some interest because, unlike ordinary resonators, my circuit is completely devoid of capacitance. Perhaps I will decide one day to check my theoretical deductions experimentally. The experiment should not be too difficult, but before I do this I would like to make a long series of studies, all of the same kind, in order to publish them as a block; and I think that to finish them the way I would like to, will require about a year of work.

If you remember, the last time we met I told you of some studies on the variation of the refractive index as a function of wavelength which I wanted to do and then dropped unfinished. I would now like to coordinate these studies, which are all based on the same principle, and obtain from them a piece of work as fully organized as possible. This is especially desirable because in the mathematical

Letters to Enrico Persico

handling of each of these problems one encounters just about the same questions, and hence they are suitable for a unique approach. I suggest you make some studies on the kinetic constitution of matter (in this connection, the Brownian motions which you mention to me are movements observed with the ultramicroscope in colloidal solutions). When you have time, try to solve this problem. The variables x and y and their functions ξ and η are considered Cartesian coordinates in two planes. Then a line in the first plane corresponds to a line in the second. It is required that two arbitrary lines in the first plane intersect at an angle equal to the angle of intersection of the corresponding lines in the second plane. In this way you will have a little practice with partial derivatives, with which you were not very familiar. Please give me in return some fairly difficult problem.

In the meantime, in order not to waste my time, I am reading the *Théorie des tourbillons* of Poincaré. It is a rather interesting hydrodynamical theory, to which great importance has been attributed for explaining the constitution of matter by introducing the usual hypothetical fluid. Enough! It is better that I don't say any more; otherwise I would say too much.

As a pastime I have restudied the question of the Amsler planimeter and have been able to solve the problem of finding how many turns the well-known little wheel completes when the pen of the instrument moves along an open curve. If you remember, when the curve is closed this number is proportional to the area. I have also solved the problem in the case in which the pen describes a closed curve containing in its interior the fixed point of the instrument. If you have time, you might try to study this problem, which is fairly instructive from the point of view of kinematics. Solve many problems of rational mechanics, the experience will be very helpful to you in improving your knowledge of mathematics. You can look in Appell,[7] for instance, where you will find as many problems as you could possibly want.

At this time, since I have almost nothing to do for school, and with many books available, I am trying gradually to enlarge my knowledge of mathematical physics, and I will try to do the same for pure mathematics also, because the farther I go, the more I find they are both necessary to me. Furthermore, in studying one of these two things one learns the other also, and I have certainly learned more mathematics in physics books than in those of mathematics. If you can find it, it will be very good for you to look at the treatise on

Letters to Enrico Persico

rational mechanics by Poisson,[8] because, although it is a little old, it is still excellent, and you can learn many useful things from it.

Here in Pisa, too, we have an anticyclone which delights us with an intense cold. As a compensation the weather is excellent. Excuse me if I have written you too scientific a letter.

Respectful greetings to your parents, and to you a handshake.

Your friend,
Enrico Fermi

Pisa, June 8, 1919

Dear Enrico,

Please don't be too hopeful about my conversion to music, and don't ask too much of me.

As to your circular billiard table,[9] from a cursory examination of the problem it seems to me that it leads to an equation of the fourth degree. I am not sure of this but, in any case the problem is equivalent to the other one of tracing an ellipse given the foci and a circle tangent to it.

I hear that you have devoted yourself to the study of partial differential equations. I suggest that if you have the time, you consult a chapter in the *Mécanique* of Poisson[10]—contained in the section on dynamics of continuous systems—in which he explains a method I mentioned to you on another occasion and which in many cases, even if it does not give the complete solution of the problem, at least allows one to determine the periods for small oscillations—the question which in most cases is of the greatest practical interest.

As for our summer activities, I would still like to try determining the capillary constant of water, and I am trying to find methods that might make this measurement reasonably easy. Among others, I have thought of using the measurement of the oscillation period of a water drop; this would not be very difficult from the experimental point of view. I remember an experiment of this type which is described in the book by Murani;[11] there are difficulties on the theoretical side, however, and they are rather serious. If I succeed in overcoming them, I think the method is such that it could give satisfactory results. In the meantime I am actively continuing to reorganize my notions of physics—an enterprise which succeeds fairly easily—using those few and very disorderly notes that I still preserve. I have also read almost all of Appell,[12] especially the parts that treat the general

Letters to Enrico Persico

theorems of motion and the consequences of the Hamilton-Lagrange-Jacobi equations. I regret that I have not succeeded in finding in the library the third part of this work, which treats the dynamics of continuous systems. Anyway, we will speak about all this in person with more leisure because I hope to arrive in Rome not later than the end of this month.

Regards to your parents, greetings, and best wishes.

Affectionately yours,
Enrico Fermi

Pisa, December 11, 1919

Dear Enrico,

Finally I have decided to study chemistry, theoretical and nontheoretical. I'm studying the theoretical from Nerst's [sic] *Theoretische Chemie* (760 pages) and the nontheoretical, for the time being, from Ostwald's *Foundations of Inorganic Chemistry* (818 pages); I will decide later what to do for organic chemistry. The lectures, for the time being, do not keep me very busy—eight hours a week. To these I add six hours a day of chemistry plus some physics lab, variable from day to day. I hope that, proceeding at this speed, I will be able by the end of winter to free myself of the beloved chemistry and not have to think about it for a long time. After this I will busy myself with those investigations,[13] and finally I shall start to think about a subject for my doctoral dissertation. And what are you doing? Have you decided which courses to take? Have you finished reading Planck's *Thermodynamik?*

This year among other things, I have dedicated myself to the publishing of lecture notes; it seems that this business thrives in spite of the competition from three freshmen, who apparently have now been forced into bankruptcy. We shall see how things go. The course notes are for experimental physics (mechanics and perhaps heat and optics); I dictate them, and another boy takes care of writing and printing them. The earnings are divided equally. It is not unlikely that, if I have time, I will dedicate myself to the construction of a precision balance (I would like at least one-milligram sensitivity) with materials and instruments of the physics laboratory. Anyway, I will start to work before the Christmas vacation, which I think starts Sunday the twentieth.

Letters to Enrico Persico

Thus we will soon have occasion to meet again in Rome. In the meantime, kind regards to your parents and greetings to you.

Affectionately,
Enrico Fermi

Pisa, January 30, 1920

I write two words to you in a hurry, just to remind you of my existence. In a few days we will talk to each other, when I come to Rome for the *carnevale* vacation from the twelfth to the eighteenth of February.

My studies are proceeding very well because I have almost freed myself of the inorganic chemistry, and I have decided to study organic chemistry in class. We have four hours a week of organic chemistry alone, and I hope this will suffice.

Thus I have again started to study the progress made by physics during the war and have found that in fact something new, not too much, has appeared. In the physics department I am slowly becoming the most influential authority. In fact, one of these days I shall hold (in the presence of several magnates) a lecture on quantum theory, of which I'm always a great propagandist, especially in relation to the phenomena of *Feinstructur* [sic], multiplicity of spectral lines of Balmer's series, and similar subjects, and also on the so-called Stark-Lo Surdo effect, which is the splitting of the hydrogen spectral lines under the action of an electric field. Both these phenomena find an incomplete explanation, which does not correspond to experiment, in the electrodynamics of Maxwell and company, however, they can be calculated in a most satisfactory quantitative way by using quantum theory.

Goodbye now; best regards to your family, and greetings to you.

Enrico Fermi

Pisa, May 30, 1920

Dear Enrico,

I have received news from home that this year you too are planning to come to Ladispoli. I am very glad about this news because we will thus have time to stay a while together. I have been very busy. I needed to measure the ratio between an inductance coefficient and a capacitance in the laboratory, and since I made a slight error in the order of magnitude of only 10^{18}, I did not succeed in orienting myself, and I have wasted more than a week to obtain, at the end, a result with three significant figures, of which I would not guarantee the last. On the other hand, starting from ideas similar to those I mentioned to you last Easter, I have tried to find a relationship (which I have finally found and which I will compare with experiments) between the width of spectral lines and damping coefficient of their oscillations. My theory, as far as I can tell, seems to explain some peculiarities of the phenomena which disagreed with the old theory of Lord Rayleigh, who ascribed the line width to the Doppler effect and to collisions between molecules.

Some time ago you spoke to me about a recent experiment in which somebody should have discovered a sort of Stark effect produced by gravitational fields. I would be very grateful to you if you were to send me the journal reference, because I have not succeeded in locating it. I have almost abandoned the idea of the photoelectric effect in gases for my dissertation. It would not be impossible instead, that I should work on the interesting phenomena of diffraction of Roentgen-rays in crystals—especially as I hope to be able to connect them easily to statistical theory, because I believe that in Roentgen rays the differences from the ordinary wave theory of light should appear much more markedly.

Affectionate greetings, and kind regards to your parents.

Enrico Fermi

Pisa, November 29, 1920

My dear Friend,[14]

I have been in Pisa for more than twenty days and have already started my work on Roentgen crystallography. The first step was to protect myself and my collaborators from the action of the X-rays.

I have achieved this by enclosing the X-ray tube in a lead box of about 3 mm thickness. It was not an easy job to build the high-tension line from the induction coil to the tube. I had to protect several parts of the line with two or three glass tubes in order to avoid sparks between the high-tension line and the lead box.

I have used a large induction coil which gives about 40 cm of spark and has an electrolytic switch (about five hundred interruptions per second); then I tried to obtain some Laue photographs. I have tried with $CaSO_4 + 2H_2O(001)$, with $SiO_2(2120)$, and finally with $CaF_2(111)$. With the first two substances I have obtained only the undeflected ray; with calcium fluoride I have made two pictures, and in both I have obtained a photogram as in the figure.

It is very remarkable that I have not obtained the symmetry to be expected from the holohedral class of the monometric system. I will investigate this anomaly further.

I am expecting news about your studies. For your dissertation I thought of the following subject: "Studies on the Number of Molecules Forming an Ion as a Function of Temperature and Pressure." I think that one could solve such a problem by plotting the curve of mobility versus pressure and noting its kinks, from very low pressure up to the pressure one wants to investigate.

My best greetings to you and kind regards to your parents.

Heinrich Fermi

Pisa, November 24, 1921

Dear Enrico,

From your letter of the twenty-third, I learned of the happy event of your beatification.[15] The illustration on the side—I don't know whether you will be able to interpret it because true works of art are always difficult to understand—shows what I would do if I were in Rome.[16] Concerning the *Nuovo Cimento,* I will take care of the summaries of the *Annalen der Physik.* Here, as usual, they make little sense. I have every intention of summarizing only those papers which have something to say, neglecting all the more or less academic exercises, of which the scientific journals are full.

I am attending courses on higher analysis (differential geometry and theory of differential equations) and higher mechanics (analytical dynamics and general methods of celestial mechanics). My thesis proceeds slowly, almost backwards.[17] I will write to you at greater length as soon as I have time.

Together with this letter I am sending you the reprints of my scientific papers.

Affectionate greetings to you and regards to your parents.

Enrico Fermi

Letters to Enrico Persico

Pisa, January 25, 1922

Dear Enrico,

I received your letter of the twentieth a few days ago. A thousand thanks also from the interested party[18] for the immediate and complete information. I am acting as lecturer, relativist, and physicist. In the first activity, as you probably have heard, I have not yet sullied my conscience with manslaughter. The philosophers have become terribly angry with me because, in their own words, "since the bases of the theory of relativity are not only physical but also logical, it would be advisable to put oneself in touch also with the results which this logic has attained." But *"Wer 'fregiert' sich darum"*:[19] until they hit me they don't scare me.

As a relativist I am trying with great effort to launch the business of the 4/3. The main difficulty derives from the fact that they have a hard time understanding—in part because the thing is not easy to understand, in part because I express myself too concisely—but little by little they begin to understand what it is all about.

According to what has been decided, I should publish this study as a memoir in the proceedings of the *Accademia dei Lincei,* in the *Nuovo Cimento,* and in the *Annalen der Physik.*

Puccianti, however, out of political considerations, which perhaps one should heed, does not want me to send it to the *Annalen* before having presented it to the *Lincei.*

As a physicist, my principal activity consists in doing nothing, because after all I think that Boltzmann statistics do not absolutely exclude the possibility that my dissertation could produce itself by thermal agitation—although such a possibility does not seem very probable. However, at a certain point I will just have to settle down and work at it seriously, because I would like to finish the experimental part around Easter. Let's put our hope in thermal agitation.

As far as the shoe-nails are concerned, I think it best to order them from Bitteland,[20] and I have given my parents full power to deal with you in this matter. For me, at least, you should order rather small nails because my shoes are not made for large ones.

As to the grid for catching ions, I have never heard that such a scheme has ever been tried, and personally I think that it might get good results.[21] The experiments on gases, however, seem to me exceedingly difficult, and I, at least, would be scared by them.

Letters to Enrico Persico

During the last few days I have been rather busy writing down my lecture on relativity, which I might publish in *Scientia*. If I ever were to do it, I want to declare to you that I do it for purely practical reasons and not because I consider it useful for the advancement of science to print articles of this kind. By the way, I have not yet definitely decided because I am a little afraid that the philosophers might rebel in fury and draw me into fruitless polemics. Enough of that; we will see.

Besides these activities, I go on my traditional Sunday bicycle trips. Usually I go with the girls of the feminine section of the SAP,[22] and furthermore I occasionally go on some AP[23] expeditions just to break the monotony of everyday life.

Affectionate greetings to you and your family and, again, many thanks.

Caeterum ego censeo Hamiltonianum principium esse applicandum.

Enrico Fermi

Pisa, March 18, 1922

Dear Enrico:

From your letter of the thirteenth I received the full news of your father's health, and you can imagine how much it grieves me, even though I had already heard a good part of it through my own family. Let's hope that the X-ray treatment has good effects and that the improvement is permanent.

I have been and still am exceedingly busy, partly because of my dissertation, which by the way has become a first-class mess.

Essentially it will consist of the following parts: an introduction, with a history and a review of the present state of the question; a theoretical part, consisting of some studies on the resolving power of reflection by very thin curved crystals and a complete study of the effect of the thermal motions on X-ray reflection; and an experimental part, consisting of obtaining a photographic image of the anticathode "according to Lockyer," by means of reflection on a curved mica sheet.

As you see, the program is very modest, but it has the advantage of being nearly finished. Certainly it will be finished before the Easter vacation, and only the writing will remain. In addition to this work I have devoted myself to *Quantentheoretische* (sic) *Betrachtungen,* which up to now have brought me to a justification of the blackbody formula from the point of view of Bohr's theory; they could perhaps carry me much further if it were not for the almost insurmountable difficulties caused by the extremely complicated calculations required. The fundamental idea of this investigation is to consider the atom and the electromagnetic field to which it is coupled as a single system and to calculate the *statischen Bahnen* of this system as a whole. The maximum program would be to remove all the incompleteness of Bohr's theory. I expect, however, that because of the difficulties I mentioned above I will not be able to conclude the investigations. So far as the abstracts for *Nuovo Cimento* are concerned, such an absurd and calorimetric [sic] criterion has been adopted that I have almost completely lost interest in them.

You have done well to free yourself of the nuisance of C. F.[24] I would have prevented this if I had had any indication from him that he would bother you. I will try to find out why you do not receive *Nuovo Cimento*.

I cannot give you any bibliographical indication on the problem of electromagnetic waves in general relativity. To my knowledge, the only problem of this type that has been studied is the propagation of waves in a metric given a priori (deflection of light rays in the vicinity of the sun). I suggest, however, that before you waste too much time on such a problem you talk to Levi-Civita and get all the information you can from him.

In a couple of hours I will leave for a trip in the mountains. But please do not mention this to my parents, if you see them before Thursday, so that they will not worry about my falling victim to some strange curvature of the *weite Welt*.

And now my best greetings and again my warmest wishes for your father's health.

Enrico Fermi
March 20, 1922

I forgot to mail this letter before leaving; now I am back. The trip went very well, and the curvature of the *Welt* did not bother me at all.

Again my greetings.

Letters to Enrico Persico

Pisa, May 25, 1922

Dear Enrico,

I have just received news of you from your letter of the twentieth and from Tieri, who, as you know, has been here as a member of the committee for Polvani's *libera docenza*. I have received the reprint of your paper "On the slow motion . . . ," for which I thank you. I will give you reprints of my two papers "On a discrepancy . . ." as soon as I receive them. I hope this will be soon. I have also sent the same paper to the *Physikalische Zeitschrift*,[25] and they tell me they will publish it soon.

I have almost finished writing my dissertation for the university. It consists of a monographic chapter on X-rays in general, which I will later publish in *Nuovo Cimento;* two theoretical chapters on the peculiarities of reflection by crystals; and finally, an experimental chapter on obtaining X-ray images by reflection from a curved crystal.[26]

As to the dissertation for the Scuola Normale, I have had the following accident. As I probably have written to you, the dissertation was to be composed of two probability theorems, both dealing with the sum of many quantities (for each of which the statistical distribution is given) and of some applications of these results. Now —a few days ago—I found out that one of these theorems is not new. As a consequence I do not know whether to extend the other theorem and present it alone or to abandon the idea of completing the dissertation for the July session. The trouble, more than just the lack of subject matter, is that to stretch the subject to the dimensions suitable for such a stupid work as a dissertation, I will be forced to add a fair amount of distilled water, and you well know my dislike of such a procedure. I shall see later.[27]

Scientifically, the only noteworthy paper I have recently seen is one in the last issue of the *Philosophical Magazine,* by somebody who shows that if an alpha-particle bombardment produces in an atom an instability similar to that of radioactive substances, the transformation period of the product must be extremely short.

I hope to see you soon. Best greetings, and regards to your mother. Send me news of yourself frequently.

Enrico

Letters to Enrico Persico

Pisa, June 2, 1922

Dear Enrico:

I am sorry to bother you with an importunity, but I do not know whom to turn to otherwise. You know that in my dissertation for the Scuola Normale I solved the following problem. Consider a comet whose orbit intercepts Jupiter's orbit. Every time the comet passes near Jupiter it is strongly deflected, and after a certain number of deflections it may happen that the orbit becomes hyperbolic and the comet disappears into space. One can prove, as seems almost obvious, that after a sufficiently long time the comet's orbit will certainly become hyperbolic if various inequalities are valid. I have calculated the probability of this happening before a given time and the probability that before this happens the comet might be destroyed by a collision with Jupiter.

Would you please have a look at the bibliographical index of Ouzau (?) [sic][28] to see if this problem has been studied? If you can, do it rather fast, because on your answer depends the handling of my thesis.[29] You could give the answer to Armellini,[30] who will be in Pisa next Tuesday.

Many, many thanks, and forgive me for bothering you.
Affectionate greetings, and regards to your mother.

Enrico

Pisa, June 8, 1922

Dear Enrico,

Many thanks for your immediate help with the comet business. I had already looked into Tisserand[31] where I could not find anything relating to the particular problem I have studied. I had asked you, at Armellini's suggestion, to look at Ouzau [sic], which is a treatise on celestial mechanics containing practically everything, but I know just about as much as you. Anyway, my paper is ready, and as soon as Armellini answers me favorably I will deliver it. My dissertation now is ready and I will hand it in in two or three days. By the day after tomorrow we will have finished all the courses and we will not speak about them anymore. I will have to pass the ex-

Letters to Enrico Persico

amination in higher analysis (differential geometry) which is a terrific bore, in which the problems studied are chosen by the sole criterion that they should lack all interest. I have, besides, an examination in the physics laboratory and one in freehand drawing (have a look at the enclosed document of my pictoral art[32]). Maybe, just for fun, I will take the examination in higher mathematics too.

One can see that when Mohs made his empirical hardness scale, science was still rather undeveloped; otherwise he surely would have chosen as follows: topaz, corundum, diamond, and the brains of female candidates for the mixed degree in mathematics and physics.

Act I. *Scene from real life.*

I, assuming professorial airs; and six young girls, barring one or two exceptions ugly enough to scare anybody. They scrutinize in a rather suspicious manner a micrometric screw. A profound problem has presented itself: reducing the fraction 1,000/200 to its lowest terms.

I: Go ahead, ladies; please reduce 1,000/200 to its simplest form
The girls, as those who agree,[33] *keep silent and smile in embarrassment.*
First Girl: Well. . . .
Second Girl: Just about. . . .
Three and Four: Yes. . . .
I: You mean you are not even able to simplify a fraction?
  (*addressing the first girl*) Be good and try.
First girl blushes modestly.
Second Girl: (*in a sudden inspiration which transfigures her unattractive face*) You divide both terms by a number different from zero.
I: (*hiding my disappointment*) By which number?
The girls: look at each other, *terrified.*
I: By their greatest common div. . . .
The girls: (*with great emphasis*) . . . isor.

I forgo, in charity to the reader, the search for the greatest common divis—or through resolution into prime factors and division of "both terms of the fraction" by it. If the reader is interested, I communicate the result of the investigation: $1{,}000/200 = 5/1 = 5$.

As you see, it is pretty bad; and those people in a few years will

Letters to Enrico Persico

be in charge of teaching mathematics (and what is worse, physics). The discussion of my dissertation will be on July 3 or 4. Kindest regards and many thanks again; remember me to your mother.

Enrico

Rome, August 8, 1922

Dear Enrico,

Yesterday I received your letter of the twenty fifth, with rather *spöttisch* remarks about Antignano and its slow-moving inhabitants. We, on the other hand, enjoyed S. Donato very much. It is a rather solitary little place, composed of the house in which we lived (which was formerly a convent), a chapel not in use, and a shack where a pastor lives with his ten children.[34] All this is a half-hour's walk from the next inhabited place. Since the company—though rather afflicted with chronic theologiphilia—was most congenial and pleasant, we spent a really charming week and left with heavy hearts to return to the heat and competitions of Rome.

As to the competitions (God be praised), last Saturday I took to the ministry the entire, heavy file of my documents—consisting of neither one nor two, nor three, nor four, nor five, neither six, nor seven, nor eight, nor nine, nor ten, but eleven publications, among which is the one I concocted while I was at S. Donato. It deals with the behavior of elastic bodies according to general relativity. In it I obtained some curious results, which I want to talk to you about. By the way, I think you will be back in Rome shortly.

Since I did not receive a reply to the postcard I wrote to Signora Rasetti some time ago about the well-known project,[35] I wrote again today to the son, hoping to reach him directly because it is now time to make a definite plan. In the meantime I will start putting my bicycle in shape so that it does not cause worries of any kind. This will be a rather big job but I think I have all the materials to do it pretty well.

I do not think I will know the results of the competition for a couple of months, because the committee will not meet before the end of September, the gentlemen of the committee having no desire to enjoy the heat of Rome. In any case, I will keep in touch with some big shot or other so as to know something as soon as possible, since my next decisions depend on the result.[36]

Cordial greetings to you and your mother from all of us, and hoping to see you soon.

Enrico Fermi

Göttingen, March 31, 1923

Dearest friend,

I returned to my third fatherland about one week ago. My trip to Berlin, Dresden, and Leipzig was interesting and pleasant, especially since in Berlin I found acquaintances of Maria[37] who kept me company.

And now I have returned to my work. I am trying to generalize the adiabatic principle to arbitrary mechanical systems. The work seems about to give fair results.

I imagine that in Rome, as usual, you will ruin your health by working too much. Take care of yourself!

I hope that by now nice weather will allow you to continue the studies with the reflector. Here, at least, we are having sun and weather unworthy of the fifty-second parallel.

Cordial greetings and good wishes, though a little late, for you and your mother.

Enrico

Göttingen, April 24, 1923

Dear Enrico,

Carrara[38] has just written me that the University of Pisa has opened a competition in higher mechanics.

Although the probability of winning is rather small, because as usual there will be applicants with twenty or more years' experience, I want to try in any case—the more so since I have three rather important publications ready which might be considered as belonging to higher mechanics. I hope they arrive in time.

By the way, I would like you to ask if there would be space for two or three of my publications, about thirty pages altogether, at the Lincei. I write you not only to tell you of my plan but also to

Letters to Enrico Persico

ask you to spread about some rumor of it—so that you may hear whether people consider the idea of my taking part in the competition too crazy.

Affectionate greetings to you and our friends at the institute. Remember me to your mother.

Enrico

Moena [Dolomites], July 26, 1924

Dear Enrico,

Many thanks for your letter, which I received in perfect order together with the passport.

I waited a few days to answer you because I have to ask you something, and precisely:
1. Range of the α particle of RaC in helium.
2. K and L energy levels for oxygen and nitrogen.

The first piece of information you will find in Rutherford's book, and the second, at least by extrapolation, in a paper by Bohr and Coster in the *Zeitschrift für Phys.* of 1923. In addition, I need the ionization potential of helium.

Forgive my bothering you, and many thanks in advance. Please tell Trabacchi that when we leave our house there is nobody, so far as we know, who seems to be interested in it.[39]

We are enjoying ourselves here very much, and the place is not inferior to our expectations.

Everybody sends you greetings, and we all wish you lots of fun at Marina di Massa.

Affectionate greetings to you and your mother.

Enrico

Leyden, September 13, 1924

Dear Enrico,

Yesterday, finally, I reached my destination after a rather long and therefore somewhat boring trip—however, without incident.

I had to postpone my departure because of an abscess behind my ear. I had to have it lanced because it did not want to disappear of itself.

Could you please—when you have a chance to go to the Minerva[40]—inquire about my *libera docenza*? For instance, have you already received the gracious invitation to pay?[41] I have not heard anything else about it. As you probably know, the job at the University of Florence can now be considered an accomplished fact, and I will go there in December.

I went to the physics laboratory today, but since they have established the English Saturday here, I was only able to see Crommelin.[42]

Next Monday they will commence classes at the university.

Affectionate greetings to you and your mother.

Enrico

Florence, May 22, 1925

Dear Enrico,

I received your letter of the sixteenth. On the whole, everybody here is rather reluctant to request the competition without previously knowing the regulations which will govern it. I think, therefore, it is practically impossible that the faculty will decide to request a competition at this point, while according to Gentile's laws[43] it should be called by May 31 to have validity for the coming year. I myself do not think it improbable that, together with the new rules for the competitions which everybody here believes might be published anytime now, there might also appear temporary dispositions that would postpone the May 31 deadline. In any case, I recommend that you try to find out what the new rules are as soon as possible, not only so that you can behave accordingly, but also because the earlier we know the rules, the greater the probability of acting successfully.

One solution which here would be looked upon more favorably than opening a competition would be that you compete for the chair in theoretical physics at Rome; and in case you place second, they would appoint you here, if the rules allow it. Here it is generally expected that they will.

Following Tricomi's[44] advice, I thought it unneccessary to state—for the time being—that you are not inclined to accept the simple "assignment." In Tricomi's opinion, the danger would be that, if the university fails to find a temporary solution for the coming year, the faculty might decide to fill the chair by transferring somebody. This would be the worst solution, both for you and for the University of Florence. Naturally, after May 31 I might communicate to them that you are unwilling to accept the assignment, because by then there would not be time for the university to ask for a transfer.

As to the competition in theoretical physics, I heard from Levi-Civita that the faculty in Rome asked for it twenty days ago. I suspect, however, that publication of the announcement in the bulletin has been delayed, as usual, awaiting the new rules. As soon as I hear something definite I will inform you immediately.

Rasetti and I have finished our studies on the effect of the alternating field.[45] We will send it to the Lincei in a few days. The results have been a complete confirmation, within the limits of error (really rather large), of the theoretical predictions according to classical theory—a little doctored, however, because one must choose a Larmor precession 3/2 times bigger than the normal one, as indicated by the anomalous Zeeman effect of the line 2536. We will also publish this work in the *Zeitschrift für Physik*.

It is not impossible that I will come to Rome next month for a fortnight because I have a long vacation between the end of classes, which will be finished at the beginning of next month, and the examinations, which will be held around the middle of July. Tricomi will be in Rome about ten days from now, and I think he will stay for a few days. I will ask him to look you up at the physics laboratory.

Greetings to you and your mother. Give my regards to Professor Corbino and all the Roman physicists.

Enrico

Florence, June 2, 1925

Dear Enrico,

I have not come to Rome yet and it is possible that I may not come at all, at least for the time being, because I have fallen victim, so to speak, to an occupational hazard. In fact, and very unexpectedly, I have been put on the board for the state high school examination,[46] and furthermore I was not informed of this until the last moment, so that I was unable to find a substitute. Therefore I shall be digesting 130 examinations throughout the month of July, and I will have to cut my visit to Rome to two or three days, if I can come at all. As to the possibility of a chair for you here in Florence, by now practically everybody agrees on the advisability of appointing you (changing the name of the chair from mathematical physics to theoretical physics) if you place second in the competition for Rome. The only danger in this plan is that the new regulations may not permit a competition to apply to a different chair than the one for which it was called. But there are contradictory rumors here. If the competition for Rome cannot be used for Florence, it might be possible to persuade the faculty to open the competition here as well. In any case, such a decision can be made only when something more definite is known.

So you have definitely decided to go to Cambridge. I recommend that you try to find a young and beautiful heiress with lots of pounds sterling at 147[47] and also—if possible—try not to distort your face too much trying to speak "with the contracted mouth."[48]

I hope this letter finds you still in Rome; if so, my best greetings to everybody.

Enrico Fermi

Rome, September 23, 1925

Dear Enrico,

It is not my fault that up to now I have shown no signs of life; I have been in such a brutish state of mind that I have not been able to formulate a coherent thought. In the hope that my ideas are now formed to the point that this letter might be intelligible, I have decided to write it.

First of all, here is a brief chronicle of this summer. During the whole month of July I was in Florence, a victim of the state exams. I spent the month of August, as you know, at S. Vito with the usual company. We had a good time but nothing happened worth recording here.

At the beginning of September, my sister and I went to Caprino Veronese to visit Cornelia;[49] then a few days at Viareggio at the Enriques'; then I came to Rome, where I am staying now. I will leave Saturday to spend a few days at Pozzuolo, to sponge the car off Rasetti, and I will be back in Florence at the beginning of October for the state examinations and to renew the slaughter.

I don't know if you heard that my sister has succeeded in getting a transfer to Rome to the Ginnasio Umberto.[50]

It is not equally sure that I (at least this year) will be coming to Rome, because it is highly probable that there will be no university competitions at all this year. The matter is now at this point: Gentile

Letters to Enrico Persico

would like the system of competitions to remain according to his reform. Father Gemelli[51] would prefer, instead, that the commission propose a group of three from which the interested faculty would be free to choose. Cirincione,[52] and with him the majority of the university professors, would like to return to the old system. These three forces are all applied to Minister Fedele.[53] Unfortunately, their vector sum is zero, and consequently the point of application does not move. Naturally, if this state of affairs lasts a little longer, there will not be enough time to open a competition this year. The whole thing is rather annoying, but there is absolutely nothing one can do about it.

During the past summer I have, of course, interrupted all scientific work. Only now have I begun reading to bring myself up to date with the current literature. My impression is that during the past few months there has not been much progress, in spite of the formal results on the zoology of spectroscopic terms achieved by Heisenberg. For my taste, they have begun to exaggerate their tendency to give up understanding things. I want to study now, from the theoretical as well as from the experimental point of view, the problem of the new spectral lines that appear in an electric or a magnetic field of suitable intensity. This may be a fertile field, and I feel it has not yet been explored as much as it deserves. An experiment that seems feasible to me is to measure the variation of the mean life of a metastable state, when it is put into a known electric or magnetic field, as a function of its intensity.

I am very anxious to know your impressions of the people in Cambridge and the persons with whom you become acquainted there. I got a postcard with your signature and another signature which I could not decipher with certainty. It seemed to be Ornstein, but I am not sure.[54] When you write to me please let me know who it was.

Affectionate greetings from us all.

Enrico

Florence, October 1, 1925

Dear Enrico,

Well, there is some news: the famous and never sufficiently praised rules for university competitions have come out. They are essentially as follows: The minister publishes the announcement of

Letters to Enrico Persico

the competition in the official newspaper and in the *Bulletin of Public Instruction*. Within two months of this announcement the competitors present their qualifications, and are judged by a committee of five persons, who choose at most three competitors, grading them according to merit. From these the faculty chooses either the highest one, who is then appointed immediately, or it can choose the second or third—but in this case, before the appointment can be made it is necessary that those higher in the list, in order of merit, either decline the position or are already in a tenure position or are called by another faculty.

The committee consists of one member appointed by the minister, two members appointed by other faculties, and two members appointed by the interested faculty. Other faculties may make use of the group of three after the faculty who asked for the opening of the competition has made its choice.

To these general rules, a few temporary rules have been added for the current year. Essentially, they extend the initial date for new appointments to February 1, 1926, and reduce the period for the submission of applications from two months to one. Also, I believe I understood that this year they will open only those competitions which were requested prior to May 31 of this year.

Coming down now from theory to practice, we conclude that the competition for theoretical physics will be for Rome and that, if Florence is willing, they can appoint the second in line. I believe that the selection committee will be composed as follows:

Corbino (or Levi-Civita or Volterra) This composition is almost
Garbasso[55]   obligatory because only
Cantone (or Majorana)[56]   one person can belong to
Somigliana[57]   the faculty requesting
Maggi[58]   the competition.

The competitors will probably be—besides you and me—Pontremoli, Polvani, Sbrana,[59] Carrelli,[60] and maybe somebody else less important. As soon as the announcement of the competition is available, I will advise you immediately. In the meantime I am preparing the documents to be presented.

I came back to Florence yesterday and began the state examinations today. As I write you, the victims are squeezing their unimaginative brains to prove why and how the Italian patriots became oriented toward the House of Savoy between 1848 and 1858.

Affectionate greetings,
Enrico

Letters to Enrico Persico

Florence, October 15, 1925

Dear Enrico,
One rather unpleasant piece of news: the competition in theoretical physics has been postponed. The reason is as follows: Last May, when the faculty asked for the opening of the competition, the course in theoretical physics was not explicitly mentioned in the bylaws of the university; the changes in the bylaws in which the new course is introduced have to be submitted to the *Consiglio Superiore* in November. If, as one hopes, they are accepted immediately, the competition might still be opened in time; otherwise it will have to be postponed till next year. Let's hope for the best. In the meantime some other competitions have been opened; the only one of interest to us is the one for mathematical physics at the University of Cagliari. I plan to compete because of the uncertainty of the competition in Rome; I consider it advisable to have a two-barreled gun, though the prospect of ending up in the Isles does not attract me particularly. Anyway, I think it is advisable that you compete too.

As soon as I have further news, which I hope will be better than today's, I will notify you.

Affectionate greetings,
Enrico

Florence-Arcetri, May 17, 1926

Dear Enrico,
The Faculty of Science here has asked for the competition in theoretical physics with the provision that the request will be withdrawn if the competition for Rome takes place. In this way we at least are almost sure that one of the two requests will land safely.

I would like you to answer the following questions for me with the *greatest speed*.
1. If and when the Faculty of Science in Rome has asked for the competition?
2. In case it has not asked for it yet, when is it going to do so?
3. What is the date of the request for the modification of the university bylaws establishing the chair in theoretical physics there?

The last point is important because there is a decree that establishes that changes in the bylaws cannot take place for five years if they have not been requested before January 31, 1926.

Letters to Enrico Persico

Waiting for the sun that does not seem to come out, I spin the electron.

Greetings and regards to you,
Professor Corbino,
Trabacchi,
Lo Surdo,
Nella (Mortara),
De Tivoli,
and others whose names I do not recall.

Enrico Fermi

P.S. I am expecting the answer at *great speed!*

Florence-Arcetri, June 29, 1926

Dear Enrico,
First of all my congratulations, even if belated, for the Sella prize.[61] And now let's get to the point! I have heard that in Rome they have chosen Corbino and Garbasso as representatives of the faculty in the committee for theoretical physics. I believe that in a few days there will be elections by the faculties to nominate other, *non-Roman* members. Barring contrary instructions, I would like to use what little influence I may have to get Cantone and Majorana elected by the faculties, because naturally I would not want those two mathematicians of last year's competition, whom I have put in my black book, on the committee. I would like Professor Corbino's opinion on the matter in time to act accordingly.

What are your plans for the holidays? I have not yet decided anything.

Greetings from all the Florentines to all the Romans.

Enrico Fermi

I hope to be able to start the experiment on the Stark effect of sunlight in a few days.[62]

Florence, September 7, 1926

Dear Enrico
Last night I arrived in Florence, thus ending my lazy summer vacation. As you know, I was first at S. Cristina and later with Ra-

Letters to Enrico Persico

setti. We took a trip, partly on foot and partly by car, in the region of Mount Adamello. Finally, I returned by car to Florence. I will remain here for a few days, always innocently hoping that fair weather will allow me to finish my by now venerable experiment on the Stark effect of light. After this I plan to go either to Bologna to the Congress of the Society of Sciences or, if—as I hear—the congress is postponed, directly to Rome. In any case I will be in Rome at the beginning of October. By the way, when you answer, I would like you to tell me how much truth there is in the news that the congress might be postponed, so that I can act accordingly.

I would also like you to go to the ministry to find out the number and identity of the competitors in theoretical physics. I imagine that you already know the names of the members of the committee; in any case I give them to you: Corbino, Garbasso, Cantone, Majorana, Levi-Civita. As to the last one, I cannot understand how he has been appointed to the committee in spite of belonging to the Faculty of Sciences of Rome.[63] I am a little worried that this fact may cause some trouble, but let's hope it won't. I certainly will not be the one to object.

Toward the end of September or the beginning of October, Rasetti plans to come to Rome, and he would like to meet Corbino; therefore, I would like to know if Corbino will be in Rome at that time.

Greetings to you and your mother.

Enrico F.

Appendix 2
ARTIFICIAL RADIOACTIVITY PRODUCED BY NEUTRON BOMBARDMENT

Although the problem of transmuting chemical elements into each other is much older than a satisfactory definition of the very concept of chemical elements, it is well known that the first and most important step towards its solution was made only nineteen years ago by the late Lord Rutherford who started the method of the nuclear bombardments. He showed on a few examples that, when the nucleus of light element is struck by a fast α–particle, some disintegration process of the struck nucleus occurs, as a consequence of which the α–particle remains captured inside the nucleus and a different particle, in many cases a proton, is emitted in its place. What remains at the end of the process is a nucleus different from the original one; different in general both in electric charge and in atomic weight.

The nucleus that remains as disintegration product coincides sometimes with one of the stable nuclei, known from the isotopic analysis; very often, however, this is not the case. The product nucleus is then different from all "natural" nuclei; the reason being that the product nucleus is not stable. It disintegrates further, with a mean life characteristic of the nucleus, by emission of an electron (positive or negative), until it finally reaches a stable form. The emission of electrons that follows with a lag in time the first practically instantaneous disintegration, is the so-called artificial radioactivity, and was discovered by Joliot and Irène Curie at the end of the year 1933.

These authors obtained the first cases of artificial radioactivity by bombarding boron, magnesium and aluminum with α–particles from a polonium source. They produced thus three radioactive isotopes of nitrogen, silicon and phosphorus, and succeeded also in separating chemically the activity from the bulk of the unmodified atoms of the bombarded substance.

This speech (*FP* 128), delivered in Stockholm on December 10, 1938, upon receipt of the Nobel Prize in physics, is reproduced, with permission, from *Les Prix Nobel en 1938* (Stockholm: Imprimerie Royale Norstedt and Söner, 1939), pp. 1–8. Copyright 1939 by The Nobel Foundation.

Artificial Radioactivity

The Neutron Bombardment

Immediately after these discoveries, it appeared evident that α–particles very likely did not represent the only type of bombarding projectiles for producing artificial radioactivity. I decided therefore to investigate from this point of view the effects of the bombardment with neutrons.

Compared with α–particles, the neutrons have the obvious drawback that the available neutron sources emit only a comparatively small number of neutrons. Indeed neutrons are emitted as products of nuclear reactions, whose yield is only seldom larger than 10^{-4}. This drawback is, however, compensated by the fact that neutrons, having no electric charge, can reach the nuclei of all atoms, without having to overcome the potential barrier, due to the Coulomb field that surrounds the nucleus. Furthermore, since neutrons practically do not interact with electrons, their range is very long, and the probability of a nuclear collision is correspondingly larger than in the case of the α–particle or the proton bombardment. As a matter of fact, neutrons were already known to be an efficient agent for producing some nuclear disintegrations.

As source of neutrons in these researches I used a small glass bulb containing beryllium powder and radon. With amounts of radon up to 800 millicuries such a source emits about 2.10^7 neutrons per second. This number is of course very small compared to the yield of neutrons that can be obtained from cyclotrons or from high voltage tubes. The small dimensions, the perfect steadiness and the utmost simplicity are, however, sometimes very useful features of the radon + beryllium sources.

Nuclear Reactions Produced by Neutrons

Since the first experiments I could prove that the majority of the elements tested became active under the effect of the neutron bombardment. In some cases the decay of the activity with time corresponded to a single mean life; in others to the superposition of more than one exponential decay curve.

A systematic investigation of the behaviour of the elements throughout the periodic table was carried out by myself, with the help of several collaborators, namely Amaldi, D'Agostino, Pontecorvo, Rasetti and Segrè. In most cases we performed also a chemical analysis, in order to identify the chemical element that was the carrier of the activity. For short living substances, such an analysis must be performed very quickly, in a time of the order of one minute.

The results of this first survey of the radioactivities produced by neutrons can be summarized as follows: Out of 63 elements investi-

Artificial Radioactivity

gated, 37 showed an easily detectable activity; the percentage of the activable elements did not show any marked dependence on the atomic weight of the element. Chemical analysis and other considerations, mainly based on the distribution of the isotopes, permitted further to identify the following three types of nuclear reactions giving rise to artificial radioactivity:

(1) $$^{M}_{Z}A + ^{1}_{0}n = ^{M-3}_{Z-2}A + ^{4}_{2}He$$

(2) $$^{M}_{Z}A + ^{1}_{0}n = ^{M}_{Z-1}A + ^{1}_{1}H$$

(3) $$^{M}_{Z}A + ^{1}_{0}n + ^{M+1}_{Z}A$$

where $^{M}_{Z}A$ is the symbol for an element with atomic number Z and mass number M and n is the symbol of the neutron.

The reactions of the types (1) and (2) occur chiefly among the light elements, while those of the type (3) are found very often also for heavy elements. In many cases the three processes are found at the same time in a single element. For instance neutron bombardment of aluminum that has a single isotope ^{27}Al, gives rise to three radioactive products: ^{24}Na, with a period of 15 hours by process (1); ^{27}Mg, with a period of 10 minutes by process (2); and ^{28}Al with a period of 2.3 minutes by process (3).

As mentioned before, the heavy elements usually react only according to the process (3) and therefore, but for certain complications to be discussed later, and for the case in which the original element has more than one stable isotope, they give rise to an activity decaying exponentially. A very striking exception to this behaviour is found for the activities induced by neutrons in the naturally active elements thorium and uranium. For the investigation of these elements it is necessary to purify first the element as thoroughly as possible from the daughter substances that emit β–particles. When thus purified, both thorium and uranium emit spontaneously only α–particles, that can be immediately distinguished by absorption from the β–activity induced by the neutrons.

Both elements show a rather strong induced activity when bombarded with neutrons; and in both cases the decay curve of the induced activity shows that several active bodies with different mean lives are produced. We attempted since the spring of 1934 to isolate chemically the carriers of these activities, with the result that the carriers of some of the activities of uranium are neither isotopes of uranium itself, nor of the elements lighter than uranium down to the atomic number 86. We concluded that the carrier was one or

Artificial Radioactivity

more elements of atomic number larger than 92; we use to call the elements 93 and 94 in Rome with the names of Ausonium and Hesperium respectively. It is known that O. Hahn and L. Meitner have investigated very carefully and extensively the decay products of irradiated uranium, and were able to trace among them elements up to the atomic number 96*.

It should be noticed here, that besides the processes (1), (2) and (3) for the production of artificial radioactivity with neutrons, neutrons of sufficiently high energy can react also as follows, as was first shown by Heyn: The primary neutron does not remain bound in the nucleus, but knocks off instead one of the nuclear neutrons out of the nucleus; the result is a new nucleus, that is isotopic with the original one and has atomic weight less by one unit. The final result is therefore identical with the products obtained by means of the nuclear photoeffect (Bothe), or by bombardment with fast deuterons. One of the most important results of the comparison of the active products obtained by these processes, is the proof, first given by Bothe, of the existence of isomeric nuclei, analogous to the isomers UX_2 and UZ recognized long since by O. Hahn in his researches on the uranium family. The number of well established cases of isomerism appears to increase rather rapidly, as investigation goes on and represents an attractive field of research.

THE SLOW NEUTRONS

The intensity of the activation as a function of the distance from the neutron source shows in some cases anomalies apparently dependent on the objects that surround the source. A careful investigation of these effects led to the unexpected result that surrounding both source and body to be activated with masses of paraffin, increases in some cases the intensity of activation by a very large factor (up to 100). A similar effect is produced by water, and in general by substances containing a large concentration of hydrogen. Substances not containing hydrogen show sometimes similar features, though extremely less pronounced.

The interpretation of these results was the following. The neutron and the proton having approximately the same mass, any elastic impact of a fast neutron against a proton initially at rest, gives rise to a partition of the available kinetic energy between neutron and proton; it can be shown that a neutron having an initial energy of

---

* The discovery by Hahn and Strassmann of barium among the disintegration products of bombarded uranium, as a consequence of a process in which uranium splits into two approximately equal parts, makes it necessary to reexamine all the problems of the transuranic elements, as many of them might be found to be products of a splitting of uranium.

217

Artificial Radioactivity

10^6 volts, after about 20 impacts against hydrogen atoms has its energy already reduced to a value close to that corresponding to thermal agitation. It follows that, when neutrons of high energy are shot by a source inside a large mass of paraffin or water, they very rapidly lose most of their energy and are transformed into "slow neutrons." Both theory and experiment show that certain types of neutron reactions, and especially those of type (3), occur with a much larger cross section for slow neutrons than for fast neutrons, thus accounting for the larger intensities of activation, observed when irradiation is performed inside a large mass of paraffin or water.

It should be remarked furthermore that the mean free path for the elastic collisions of neutrons against hydrogen atoms in paraffin, decreases rather pronouncedly with the energy. When therefore, after three or four impacts, the energy of the neutron is already considerably reduced, its probability of diffusing outside the paraffin, before the process of slowing down is completed, becomes very small.

To the large cross section for the capture of slow neutrons by several atoms, there must obviously correspond a very strong absorption of these atoms for the slow neutrons. We investigated systematically such absorptions, and found that the behaviour of different elements in this respect is widely different; the cross section for the capture of slow neutrons varies, with no apparent regularity for different elements from about 10^{-24} cm^2 or less, to about a thousand times as much. Before discussing this point, as well as the dependence of the capture cross section on the energy of the neutrons, we shall first consider how far down the energy of the primary neutrons can be reduced by the collisions against the protons.

THE THERMAL NEUTRONS

If the neutrons could go on indefinitely diffusing inside the paraffin, their energy would evidently reach finally a mean value equal to that of thermal agitation. It is possible, however, that, before the neutrons have reached this lowest limit of energy, either they escape by diffusion out of the paraffin, or are captured by some nucleus. If the neutron energy reaches the thermal value, one should expect the intensity of the activation by slow neutrons to depend upon the temperature of the paraffin.

Soon after the discovery of the slow neutrons, we attempted to find a temperature dependence of the activation, but, owing to insufficient accuracy, did not succeed. That the activation intensities depend upon the temperature was proved some months later by Moon and Tillman in London; as they showed, there is a considerable increase in the activation of several detectors, when the paraffin, in which the neutrons are slowed down, is cooled from room temperature to liquid air temperature. This experiment definitely proves that a considerable percentage of the neutrons actually reaches the energy of thermal agitation. Another consequence is that the diffusion process must go on inside the paraffin for a relatively long time.

In order to measure directly at least the order of magnitude of this time, an experiment was attempted by myself and my collaborators. The source of neutrons was fastened at the edge of a rotating wheel, and two identical detectors were placed on the same edge, at equal distances from the source, one in front and one behind with respect to the sense of rotation. The wheel was then spun at a very high speed inside a fissure in a large paraffin block. We found that, while, with the wheel at rest, the two detectors became equally active, when the wheel was in motion during the activation, the detector that was behind the source became considerably more active than the one in front. From a discussion of this experiment was deduced, that the neutrons remain inside the paraffin for a time of the order of 10^{-4} seconds.

Other mechanical experiments with different arrangements were performed in several laboratories. For instance Dunning, Fink, Mitchell, Pegram and Segrè in New York, built a mechanical velocity selector, and proved by direct measurement, that a large amount of the neutrons diffusing outside of a block of paraffin, have actually a velocity corresponding to thermal agitation.

After their energy is reduced to a value corresponding to thermal agitation, the neutrons go on diffusing without further change of their average energy. The investigation of this diffusion process, by Amaldi and myself, showed that thermal neutrons in paraffin or water can diffuse for a number of paths of the order of 100 before being captured. Since, however, the mean free path of the thermal neutrons in paraffin is very short (about 0.3 cm) the total displacement of the thermal neutrons during this diffusion process is rather small (of the order of 2 or 3 cm). The diffusion ends when the thermal neutron is captured, generally by one of the protons, with production of a deuteron. The order of magnitude for this capture probability can be calculated, in good agreement with the experimental value, on the assumption that the transition from a free neutron state to the state in which the neutron is bound in the deuteron is due to the magnetic dipole moments of the proton and the neutron. The binding energy set free in this process, is emitted in the form of γ-rays first observed by Lea.

All the processes of capture of slow neutrons by any nucleus are generally accompanied by the emission of γ-rays. Immediately after

Artificial Radioactivity

the capture of the neutron, the nucleus remains in a state of high excitation and emits one or more γ-quanta, before reaching the fundamental state. γ-rays emitted by this process were investigated by Rasetti and by Fleischmann.

Absorption Anomalies

A theoretical discussion of the probability of capture of a neutron by a nucleus, under the assumption, that the energy of the neutron is small compared with the differences between neighbouring energy levels in the nucleus, leads to the result that the cross section for the capture process should be inversely proportional to the velocity of the neutron. While the result is in qualitative agreement with the high efficiency of the slow neutron bombardment, observed experimentally, it fails on the other hand to account for several features of the absorption process, that we are now going to discuss.

If the capture probability of a neutron were inversely proportional to its velocity, one would expect two different elements to behave in exactly the same way as absorbers of the slow neutrons, provided the thicknesses of the two absorbers were conveniently chosen, so as to have equal absorption for neutrons of a given energy. That the absorption obeys instead more complicated laws, was soon observed by Moon and Tillman and other authors who showed that the absorption by a given element appears, as a rule, to be larger, when the slow neutrons are detected by means of the activity induced in the same element. That the simple law of inverse proportionality does not hold, was also proved by a direct mechanical experiment by Dunning, Pegram, Rasetti and others in New York.

In the winter of 1935-36 a systematic investigation of these phenomena was carried out by Amaldi and myself. The result was, that each absorber of the slow neutrons has one or more characteristic absorption bands, usually for energies below 100 volts. Besides this or these absorption bands, the absorption coefficient is always large also for neutrons of thermal energy. Some elements, especially cadmium, have their characteristic absorption band overlapping with the absorption in the thermal region. This element absorbs therefore very strongly the thermal neutrons, while it is almost transparent to neutrons of higher energies. A thin cadmium sheet is therefore used for filtering the thermal neutrons out of the complex radiation that comes out of a paraffin block containing inside a neutron source.

Bohr and Breit and Wigner proposed independently to explain the above anomalies as due to resonance with a virtual energy level of the compound nucleus (i.e. the nucleus composed by the bombarded nucleus and the neutron). Bohr went much farther in giving also a qualitative explanation of the large probability for the exis-

Artificial Radioactivity

tence of at least one such level, within an energy interval of the order of magnitude of 100 volts corresponding to the energy band of the slow neutrons. This band corresponds, however, to an excitation energy of the compound nucleus of many million volts, representing the binding energy of the neutron. Bohr could show that, since nuclei, and especially heavy nuclei, are systems with a very large number of degrees of freedom, the spacing between neighbouring energy levels decreases very rapidly with increasing excitation energy. An evaluation of this spacing shows that, whereas, for low excitation energies, the spacing is of the order of magnitude of 10^5 volts, for high excitation energies, of the order of ten million volts, it is reduced, for elements of mean atomic weight, to less than one volt. It is therefore a very plausible assumption that one (or more) such level lies within the slow neutron band thus explaining the large frequency of the cases in which absorption anomalies are observed.

Before concluding this review of the work on artificial radioactivity produced by neutrons, I feel it as a duty to thank all those who have contributed to the success of these researches. I must thank in particular all my collaborators that have already been mentioned; the Istituto di Sanità Pubblica in Rome and especially Prof. G. C. Trabacchi, for the supply of all the many radon sources that have been used; the Consiglio Nazionale delle Ricerche for several grants.

Appendix 3
PHYSICS AT COLUMBIA UNIVERSITY
THE GENESIS OF THE NUCLEAR ENERGY PROJECT

The following is a verbatim transcript of Enrico Fermi's last address before the American Physical Society, delivered informally and without notes at Columbia University's McMillin Theater on Saturday morning, January 30, 1954. His retiring presidential address was delivered one day earlier. The present speech, transcribed from a tape recording, is left deliberately in an unpolished and unedited form. Such informality would no doubt have been frowned upon by Fermi, who was very particular about his published writings. For those who knew Fermi or heard him speak, however, the verbatim transcript may serve (as no formal document could ever serve) to bring back for a moment the very sound of his voice. The paper was presented as part of the session "Physics at Columbia University" during the Society's 1954 annual meeting.

Mr. Chairman, Dean Pegram, fellow Members, Ladies and Gentlemen:

It seems fitting to remember, on this 200th anniversary of Columbia University, the key role that the University played in the early experimentation and the organization of the early work that led to the development of atomic energy.

I had the good fortune to be associated with the Pupin Laboratories through the period of time when at least the first phase of this development took place. I had had some difficulties in Italy and I will always be very grateful to Columbia University for having offered me a position in the Department of Physics at the most opportune moment. And in addition this offer gave me, as I said, the rare opportunity of witnessing the series of events to which I have referred.

In fact I remember very vividly the first month, January, 1939,

---

The text of the address (*FP* 269) and the prefatory note are reproduced with permission from *Physics Today* 8 (November 1955): 12–16. A tape recording and records of this speech are in possession of the American Institute of Physics.

that I started working at the Pupin Laboratories because things began happening very fast. In that period, Niels Bohr was on a lecture engagement in Princeton and I remember one afternoon Willis Lamb came back very excited and said that Bohr had leaked out great news. The great news that had leaked out was the discovery of fission and at least an outline of its interpretation; the discovery as you well remember goes back to the work of Hahn and Strassmann and at least the first idea for interpretation came through the work of Lise Meitner and Frisch who were at that time in Sweden.

Then, somewhat later that same month, there was a meeting in Washington organized by the Carnegie Institution in conjunction with George Washington University where I took part with a number of people from Columbia University and where the possible importance of the new-discovered phenomenon of fission was first discussed in semi-jocular earnest as a possible source of nuclear power. Because it was conjectured, if there is fission with a very serious upset of the nuclear structure, it is not improbable that some neutrons will be evaporated. And if some neutrons are evaporated, then they might be more than one; let's say, for the sake of argument, two. And if they are more than one, it may be that the two of them, for example, may each one cause a fission and from that one sees of course a beginning of the chain reaction machinery.

So that was one of the things that was discussed at that conference and started a small ripple of excitement about the possibility of releasing nuclear energy. At the same time experimentation was started feverishly in many laboratories, including Pupin, and I remember before leaving Washington I had a telegram from Dunning announcing the success of an experiment directed to the discovery of the fission fragments. The same experiment apparently was at the same time carried out in half a dozen places in this country and in three or four, in fact I think slightly before, in three or four places in Europe.

Now a rather long and laborious work was started at Columbia University in order to firm up these vague suggestions that had been made as to the possibilities that neutrons were emitted and to try to see whether neutrons were in fact emitted when fission took place and if so how many they would be, because clearly a matter of numbers is in this case extremely important because a little bit greater or a little bit lesser probability might have made all the difference between possibility and impossibility of a chain reaction.

Now this work was carried on at Columbia simultaneously by Zinn and Szilard on one hand and by Anderson and myself on the other hand. We worked independently and with different methods, but of course we kept close contact and we kept each other informed

Physics at Columbia University

of the results. At the same time the same work was being carried out in France by a group headed by Joliot and von Halban. And all the three groups arrived at the same conclusion—I believe Joliot may be a few weeks earlier than we did at Columbia—namely that neutrons are emitted and they were rather abundant, although the quantitative measurement was still very uncertain and not too reliable.

A curious circumstance related to this phase of the work was that here for the first time secrecy that has been plaguing us for a number of years started and, contrary to perhaps what is the most common belief about secrecy, secrecy was not started by generals, was not started by security officers, but was started by physicists. And the man who is most responsible for this certainly extremely novel idea for physicists was Szilard.

I don't know how many of you know Szilard; no doubt very many of you do. He is certainly a very peculiar man, extremely intelligent (*laughter*). I see that this is an understatement. (*laughter*). He is extremely brilliant and he seems somewhat to enjoy, at least that is the impression that he gives to me, he seems to enjoy startling people.

So he proceeded to startle physicists by proposing to them that given the circumstances of the period—you see it was early 1939 and war was very much in the air—given the circumstances of that period, given the danger that atomic energy and possibly atomic weapons could become the chief tool for the Nazis to enslave the world, it was the duty of the physicists to depart from what had been the tradition of publishing significant results as soon as the *Physical Review* or other scientific journals might turn them out, and that instead one had to go easy, keep back some results until it was clear whether these results were potentially dangerous or potentially helpful to our side.

So Szilard talked to a number of people and convinced them that they had to join some sort of—I don't know whether it would be called a secret society, or what it would be called. Anyway to get together and circulate this information privately among a rather restricted group and not to publish it immediately. He sent in this vein a number of cables to Joliot in France, but he did not get a favorable response from him and Joliot published his results more or less like results in physics had been published until that day. So that the fact that neutrons are emitted in fission in some abundance—the order of magnitude of one or two or three—became a matter of general knowledge. And, of course, that made the possibility of a chain reaction appear to most physicists as a vastly more real possibility than it had until that time.

Another important phase of the work that took place at Columbia

Physics at Columbia University

University is connected with the suggestion on purely theoretical arguments, by Bohr and Wheeler, that of the two isotopes of uranium it was not the most abundant uranium 238 but it was the least abundant uranium 235, present as you know in the natural uranium mixture to the tune of 0.7 of a percent, that was responsible at least for most of the thermal fission. The argument had to do with an even number of neutrons in uranium 238 and an odd number of neutrons in uranium 235 which, according to a discussion of the binding energies that was carried out by Bohr and Wheeler, made plausible that uranium 235 should be more fissionable.

Now it clearly was very important to know the facts also experimentally and work was started in conjunction by Dunning and Booth at Columbia University and by Nier. Nier took the mass spectrographic part of this work, attempting to separate a minute but as large as possible amount of uranium 235, and Dunning and Booth at Columbia took over the part of using this minute amount in order to test whether or not it would undergo fission with a much greater cross section than ordinary uranium.

Well, you know of course by now that this experiment confirmed the theoretical suggestion of Bohr and Wheeler, indicating that the key isotope of uranium, from the point of view of any attempt of—for example—constructing a machine that would develop nuclear energy, was in fact uranium 235. Now you see the matter is important for the following reasons that at the time were appreciated perhaps less definitely than at the present moment.

The fundamental point in fabricating a chain reacting machine is of course to see to it that each fission produces a certain number of neutrons and some of these neutrons will again produce fission. If an original fission causes more than one subsequent fission then of course the reaction goes. If an original fission causes less than one subsequent fission then the reaction does not go.

Now, if you take the isolated pure isotope U^{235}, you may expect that the unavoidable losses of neutrons will be minor, and therefore if in the fission somewhat more than one neutron is emitted then it will be merely a matter of piling up enough uranium 235 to obtain a chain reacting structure. But if to each gram of uranium 235 you add some 140 grams of uranium 238 that come naturally with it, then the competition will be greater, because there will be all this ballast ready to snatch away the not too abundant neutrons that come out in the fission and therefore it was clear at the time that one of the ways to make possible the production of a chain reaction was to isolate the isotope U^{235} from the much more abundant isotope U^{238}.

Now, at present we have in our laboratories a row of bottles labeled, more or less, isotope—what shall I say—iron 56, for example,

Physics at Columbia University

or uranium 235 or uranium 238 and these bottles are not quite as common as would be a row of bottles of chemical elements, but they are perfectly easily obtainable by putting due pressure on the Oak Ridge Laboratory (*laughter*). But at that time isotopes were considered almost magically inseparable. There was to be sure one exception, namely deuterium, which was already at that time available in bottles. But of course deuterium is an isotope in which the two isotopes hydrogen one and hydrogen two have a ratio of mass one to two, which is a very great ratio. But in the case of uranium the ratio of mass is merely 235 to 238, so the difference is barely over one percent. And that, of course, makes the differences of these two objects so tiny that it was not very clear that the job of separating large amounts of uranium 235 was one that could be taken seriously.

Well, therefore, in those early years near the end of 1939 two lines of attack to the problem of atomic energy started to emerge. One was as follows. The first step should be to separate in large amounts, amounts of kilograms or maybe amounts of tens of kilograms or maybe of hundreds of kilograms, nobody really knew how much would be needed, but something perhaps in that order of magnitude, separate such at that time fantastically large-looking amounts of uranium 235 and then operate with them without the ballast of the associated much larger amounts of uranium 238. The other school of thought was predicated on the hope that perhaps the neutrons would be a little bit more and that perhaps using some little amount of ingenuity one might use them efficiently and one might perhaps be able to achieve a chain reaction without having to separate the isotopes, a task as I say that at that time looked almost beyond human possibilities.

Now I personally had worked many years with neutrons, and especially slow neutrons, so I associated myself with the second team that wanted to use nonseparated uranium and try to do the best with it. Early attempts and studies, discussions, on how to separate the isotopes of uranium were started by Dunning and Booth in close consultation with Professor Urey. On the other hand, Szilard, Zinn, Anderson, and myself started experimentation on the other line whose first step involved lots of measurements.

Now, I have never yet quite understood why our measurements in those days were so poor. I'm noticing now that the measurements that we are doing on pion physics are very poor, presumably just because we have not learned the tricks. And, of course, the facilities that we had at that time were not as powerful as they are now. It's much easier to carry out experimentation with neutrons using a pile as a source of neutrons than it was in those days using radium-beryllium sources when geometry was the essential item to control or using the cyclotron when intensity was the desired feature rather than good geometry.

Well, we soon reached the conclusion that in order to have any chance of success with natural uranium we had to use slow neutrons. So there had to be a moderator. And this moderator could have been water or other substances. Water was soon discarded; it's very effective in slowing down neutrons, but still absorbs a little bit too many of them and we could not afford that. Then it was thought that graphite might be perhaps the better bet. It's not as efficient as water in slowing down neutrons; on the other hand little enough was known of its absorption properties that the hope that the absorption might be very low was quite tenable.

This brings us to the fall of 1939 when Einstein wrote his now famous letter to President Roosevelt advising him of what was the situation in physics—what was brewing and that he thought that the government had the duty to take an interest and to help along this development. And in fact help came along to the tune of $6000 a few months after and the $6000 were used in order to buy huge amounts—or what seemed at that time when the eye of physicists had not yet been distorted—(*laughter*) what seemed at that time a huge amount of graphite.

So physicists on the seventh floor of Pupin Laboratories started looking like coal miners (*laughter*) and the wives to whom these physicists came back tired at night were wondering what was happening. We know that there is smoke in the air, but after all. . . (*laughter*).

Well, what was happening was that in those days we were trying to learn something about the absorption properties of graphite, because perhaps graphite was no good. So, we built columns of graphite, maybe four feet on the side or something like that, maybe ten feet high. It was the first time when apparatus in physics, and these graphite columns were apparatus, was so big that you could climb on top of it—and you had to climb on top of it. Well cyclotrons were the same way too, but anyway that was the first time when I started climbing on top of my equipment because it was just too tall—I'm not a tall man (*laughter*).

And the sources of neutrons were inserted at the bottom and we were studying how these neutrons were first slowed down and then diffused up the column and of course if there had been a strong absorption they would not have diffused very high. But because it turned out that the absorption was in fact small, they could diffuse quite readily up this column and by making a little bit of mathematical analysis of the situation it became possible to make the first

Physics at Columbia University

guesses as to what was the absorption cross section of graphite, a key element in deciding the possibility or not of fabricating a chain reacting unit with graphite and natural uranium.

Well, I will not go into detail of this experimentation. That lasted really quite a number of years and required really quite many hours and many days and many weeks of extremely hard work. I may mention that very early our efforts were brought in connection with similar efforts that were taking place at Princeton University where a group with Wigner, Creutz and Bob Wilson set to work making some measurements that we had no possibility of carrying out at Columbia University.

Well, as time went on, we began to identify what had to be measured and how accurately these things that I shall call "eta," f, and p—I don't think I have time to define them for you—these three quantities "eta," f, and p had to be measured to establish what could be done and what could not be done. And, in fact, if I may say so, the product of "eta," f, and p had to be greater than one. It turns out, we now know, that if one does just about the best this product can be 1.1.

So, if we had been able to measure these three quantities to the accuracy of one percent we might have found that the product was for example 1.08 plus or minus 0.03 and if that had been the case we would have said let's go ahead, or if the product had turned out to be 0.95 plus or minus 0.03 perhaps we would have said just that this line of approach is not very promising, and we had better look for something else. However I've already commented on the extremely low quality of the measurements in neutron physics that could be done at the time—where the accuracy of measuring separately either "eta," or f, or p was perhaps with a plus or minus of 20 percent (*laughter*). If you compound, by the well-known rules of statistics, three errors of 20 percent you will find something around 35 percent. So if you should find, for example, 0.9 plus or minus 0.3 —what do you know? Hardly anything at all (*laughter*). If you find 1.1 plus or minus 0.3—again, you don't know anything much. So that was the trouble and in fact if you look in our early work—what were the detailed values given by this or that experimenter to, for example, "eta" you find that it was off 20 percent and sometimes greater amounts. In fact I think it was strongly influenced by the temperament of the physicist. Shall we say optimistic physicists felt it unavoidable to push these quantities high and pessimistic physicists like myself tried to keep them somewhat on the low side (*laughter*).

Anyway, nobody really knew and we decided therefore that one had to do something else. One had to devise some kind of experiment that would give a complete over-all measurement directly of the product "eta," f, p without having to measure separately the three,

Physics at Columbia University

because then perhaps the error would sort of drop down and permit us to reach conclusions.

Well, we went to Dean Pegram, who was then the man who could carry out magic around the University, and we explained to him that we needed a big room. And when we say big we meant a really big room, perhaps he made a crack about a church not being the most suited place for a physics laboratory in his talk, but I think a church would have been just precisely what we wanted (*laughter*). Well, he scouted around the campus and we went with him to dark corridors and under various heating pipes and so on to visit possible sites for this experiment and eventually a big room, not a church, but something that might have compared in size with a church was discovered in Schermerhorn.

And there we started to construct this structure that at that time looked again in order of magnitude larger than anything that we had seen before. Actually if anybody would look at that structure now he would probably extract his magnifying glass (*laughter*) and go close to see it. But for the ideas of the time it looked really big. It was a structure of graphite bricks and spread through these graphite bricks in some sort of pattern were big cans, cubic cans, containing uranium oxide.

Now, graphite is a black substance, as you probably know. So is uranium oxide. And to handle many tons of both makes people very black. In fact it requires even strong people. And so, well we were reasonably strong, but I mean we were, after all, thinkers (*laughter*). So Dean Pegram again looked around and said that seems to be a job a little bit beyond your feeble strength, but there is a football squad at Columbia (*laughter*) that contains a dozen or so of very husky boys who take jobs by the hour just to carry them through College. Why don't you hire them?

And it was a marvelous idea; it was really a pleasure for once to direct the work of these husky boys, canning uranium—just shoving it in—handling packs of 50 or 100 pounds with the same ease as another person would have handled three or four pounds. In passing these cans fumes of all sorts of colors, mostly black, would go in the air (*laughter*).

Well, so grew what was called at the time the exponential pile. It was an exponential pile, because in the theory an exponential function enters—which is not surprising. And it was a structure that was designed to test in an integral way, without going down to fine details, whether the reactivity of the pile, the reproduction factor, would be greater or less than one. Well, it turned out to be 0.87. Now that is by 0.13 less than one and it was bad. However, at the moment we had a firm point to start from, and we had essentially to see whether we could squeeze the extra 0.13 or preferably a little

Physics at Columbia University

bit more. Now there were many obvious things that could be done. First of all, I told you these big cans were canned in tin cans, so what has the iron to do? Iron can do only harm, can absorb neutrons, and we don't want that. So, out go the cans. Then, what about the purity of the materials? We took samples of uranium, and with our physicists' lack of skill in chemical analysis, we sort of tried to find out the impurities and certainly there were impurities. We would not know what they were, but they looked impressive, at least in bulk (*laughter*). So, now, what do these impurities do?—clearly they can do only harm. Maybe they make harm to the tune of 13 percent. Finally, the graphite was quite pure for the standards of that time, when graphite manufacturers were not concerned with avoiding those special impurities that absorb neutrons. But still there was some considerable gain to be made out there, and especially Szilard at that time took extremely decisive and strong steps to try to organize the early phases of production of pure materials. Now, he did a marvelous job which later on was taken over by a more powerful organization than was Szilard himself. Although to match Szilard it takes a few able-bodied customers (*laughter*).

Well, this brings us to Pearl Harbor. At that time, in fact I believe a few days before by accident, the interest in carrying through the uranium work was spreading; work somewhat similar to what was going on at Columbia had been initiated in a number of different Universities throughout the country. And the government started taking decisive action in order to organize the work, and, of course, Pearl Harbor gave the final and very decisive impetus to this organization. And it was decided in the high councils of the government that the work on the chain reaction produced by nonseparated isotopes of uranium should go to Chicago.

That is the time when I left Columbia University, and after a few months of commuting between Chicago and New York eventually moved to Chicago to keep up the work there, and from then on, with a few notable exceptions, the work at Columbia was concentrated on the isotope-separation phase of the atomic energy project.

As I've indicated this work was initiated by Booth, Dunning, and Urey about 1940, 1939, and 1940, and with this reorganization a large laboratory was started at Columbia under the direction of Professor Urey. The work there was extremely successful and rapidly expanded into the build-up of a huge research laboratory which cooperated with the Union Carbide Company in establishing some of the separation plants at Oak Ridge. This was one of the three horses on which the directors of the atomic energy project had placed their bets, and as you know the three horses arrived almost simultaneously to the goal in the summer of 1945. I thank you. (*Applause*).

Appendix 4
THE DEVELOPMENT OF THE FIRST CHAIN-REACTING PILE

It has been known for many years that vast amounts of energy are stored in the nuclei of many atomic species and that their release is not in contradiction with the principle of the conservation of energy nor with any other of the accepted basic laws of physics. In spite of this recognized fact, it was the general opinion among physicists until recently that a large scale release of the nuclear energy would not be possible without the discovery of some new phenomenon.

The reasons for this somewhat negative attitude were the following: Two types of processes in which nuclear energy could be released might be considered in principle. Various nuclear reactions take place spontaneously with production of energy when two nuclei are brought in contact. The simplest of many possible examples is perhaps that of ordinary hydrogen. Two hydrogen nuclei when brought in contact are spontaneously capable of reacting forming a deuterium nucleus and emitting an electron. The energy liberated in this process is about 1.4Mev per process, equivalent to 1.6×10^{10} calories per gram or about two million times the combustion energy of an equal amount of coal. The reason why hydrogen is not a nuclear explosive is that two hydrogen nuclei never come in contact under ordinary conditions, owing to the repulsion of the positive electric charges of the two nuclei. There is no theoretical reason that prevents the coming together of the two nuclei; indeed this could

The American Philosophical Society of Philadelphia and the National Academy of Sciences devoted a joint meeting held in Philadelphia on November 16 and 17, 1945, to atomic energy and its implications. This paper (FP 223) was presented by Fermi at the symposium on November 17. Other speakers were: H. D. Smyth, H. C. Urey, E. P. Wigner, J. A. Wheeler for the scientific aspects; J. R. Oppenheimer for weapons; R. S. Stone for health protection; J. H. Willits, J. Viner, and A. H. Compton for the social, international, and humanistic implications; J. T. Shotwell and I. Langmuir for the problems of industrial energy.

The paper is reprinted with permission from *Proceedings of the American Philosophical Society* 90 (1946): 20–24.

First Chain-Reacting Pile

be achieved both by very high temperatures and by very high pressures. Temperatures or pressures that are, however, well beyond the limits that could be achieved by ordinary means. Actually, temperatures large enough to permit nuclear reactions to proceed at an appreciable rate are prevalent in the interior of many stars, in particular of the sun; these reactions are generally recognized as the main source of the energy irradiated by the stars.

A second possible pattern for the liberation of nuclear energy is the chain reaction. In most nuclear disintegrations particles are emitted (α–particles, protons, or neutrons) which in their turn are capable of producing new reactions. One can then conceive the possibility that when a first reaction takes place the particles produced by it may have a sufficient activity to determine in the average more than one similar reaction. When this is the case, in each "generation" the number of reacting nuclei increases until the process "burns" a sizable fraction of the original material. Whether the chain reaction develops or not depends on whether the number of new processes produced by the particles emitted by a first process is larger or smaller than one. This number is called "reproduction factor."

For all processes known until the discovery of fission early in 1939, however, the reproduction factor was in all cases enormously smaller than one. The fission process opened a new way. Almost immediately after the announcement of the discovery the possibility was discussed that when the two fragments separate they may be excited so highly that neutrons may "evaporate" out of them. This conjecture was soon confirmed by experimental observations on both sides of the Atlantic.

In the spring of 1939 it was generally known that a fission that can be produced by the collision of a single neutron with a uranium atom was capable of producing more than one new neutron, probably something of the order of two or three. It was felt at that time by many physicists that a chain reaction based on the uranium fission was a possibility well worth investigating.

At the same time this possibility was viewed with hope and with great concern. Everybody was conscious early in 1939 of the imminence of a war of annihilation. There was well founded fear that the tremendous military potentialities that were latent in the new scientific developments might be reduced to practice first by the Nazis. Nobody at that time had any basis for predicting the size of the effort that would be needed, and it well may be that civilization owes its survival to the fact that the development of atomic bombs requires an industrial effort of which no belligerent except the United States would have been capable in time of war. The political situation of the moment had a strange effect on the behavior

First Chain-Reacting Pile

of scientists. Contrary to their traditions, they set up a voluntary censorship and treated the matter as confidential long before its importance was recognized by the governments and secrecy became mandatory.

To proceed with the steps that led to the development of the chain reaction, I would like to point out that, on the basis of the information available at the end of 1939, two lines of attack to the problem appeared worthwhile. One involved as a first step the separation out of ordinary uranium of the rare isotope 235 which is responsible for the slow neutron fission of uranium. Since this separation eliminates the parasitic absorption of neutrons by the abundant isotope 238, it was felt that once uranium containing a high percentage of 235 were available it would be easy to produce a chain reaction. The real difficulty was of course to obtain isotope separation on a large scale.

The second line of attack, the one that I propose to discuss in this paper, envisaged the use of natural uranium. The problem to assemble this material in a way proper to produce a chain reaction is of course considerably more tricky than the similar problem for U^{235}. Indeed the neutrons produced by a primary fission must be used very sparingly in order to keep a positive surplus in spite of the loss due to the large parasitic absorption of U^{238}. Great care must be exerted in order to make the balance between useful and parasitic absorption of the neutrons as favorable as possible. Since the ratio of the two absorptions depends on the energy of the neutrons and, aside from details, is greater for neutrons of low energies, one of the steps consists in slowing down the neutrons from their initial high energy, which is of the order of 1 Mev, to an energy as low as that of thermal agitation. A simple process to achieve this end has been known for some time. It is based on the obvious fact that when a fast neutron collides against an atom and bounces off some of its energy is lost as recoil energy of the atom. The effect is greater for light atoms which recoil more easily and is maximum for hydrogen but quite appreciable also for all light elements.

In order to slow down the neutrons we shall have, therefore, to spread the uranium throughout a mass of some convenient light element. The most obvious choice would be the lightest element, hydrogen, which has currently been used in its combination forms of water or paraffin for the slowing down of neutrons. Further study indicates, however, that hydrogen is not well suited to the purpose. This is due to the fact that the hydrogen nuclei have an appreciable tendency to absorb neutrons with which they combine to form the heavy hydrogen nucleus, deuterium. For this reason when hydrogen is used for slowing down the neutrons a new parasitic absorption is

First Chain-Reacting Pile

introduced which eats up dangerously into the small positive excess of neutrons needed to maintain the chain reaction.

Other light elements had, therefore, to be considered for slowing down the neutrons. None of them is as effective as hydrogen for this purpose but it was hoped that their lower absorption might overcompensate for this drawback. Very little was known in 1939 of the absorption properties of many light elements. Only in a few cases rather uncertain upper limits were to be found in the literature. The most obvious choices appeared at the time to be deuterium in the form of heavy water, helium, beryllium, or carbon in the form of graphite.

In the discussions that we had in the group working on the problem at Columbia University in 1939 and 1940, and which included George Pegram, Leo Szilard, and Herbert Anderson, we reached the conclusion that graphite offered the most hopeful possibilities due primarily to the ready availability of this substance. In the spring of 1940 experimental work on the properties of graphite was initiated at Columbia University using a few tons of graphite supplied to us through the Chairman of the Uranium Committee, Dr. Briggs. Two problems were attacked and solved at that time. One consisted in the determination of the absorption properties of graphite for neutrons and one in the study of its effectiveness for slowing down neutrons. The technique used in these experiments consisted in setting up a square column of graphite a few feet thick. A small source of neutrons consisting of a few grams of beryllium mixed with radon, or radium, was placed on the axis of this column. The neutrons emitted by it diffuse through the column and are gradually slowed down to thermal agitation energy; they keep on diffusing after this until they are either absorbed or diffuse out of the column. The distribution, both in space and energy, of the neutrons throughout the column was mapped using detectors sensitive to neutrons of various energies and the results were fitted into a mathematical theory of the diffusion process. The results of these investigations permitted to develop a mathematical method for calculating with fair accuracy all the life history of a neutron from the moment of its emission as a fast neutron to the moment of its final absorption.

At the same time work was initiated in order to determine the excess number of neutrons emitted by natural uranium when a thermal neutron is absorbed by it. Since a considerable fraction of the thermal neutrons absorbed by uranium is captured by U^{238} and does not give rise to fission, this excess turns out to be fairly small and makes it therefore very essential to avoid as much as possible parasitic losses so as to end up with a positive margin that may make the chain reaction a possibility. A simple trick permits a very con-

First Chain-Reacting Pile

siderable reduction of the parasitic losses that take place while the neutron is being slowed down. Instead of spreading the uranium uniformly throughout the mass of graphite, it is more convenient to arrange it in lumps distributed in some suitable lattice configuration throughout the graphite. This device makes it less probable for a neutron to encounter a uranium atom during the slowing down process when its energy makes it particularly vulnerable to parasitic absorption.

In working out the effectiveness of this method, the group working at Columbia was very materially reinforced by the collaboration with a new research group that was set up at Princeton University. In the spring of 1941 sufficient data on the details of the process had been gathered to enable one to form a relatively clear picture of the importance of the various factors and of the best devices to be used in order to minimize the unfavorable items.

In principle it would be possible to measure with great accuracy the absorption and scattering properties of neutrons for all energies and for all atoms involved and to use these results in a mathematical theory of the process so elaborate as to make it possible to predict the behavior of a given system accurately enough to answer the question whether a given system would or would not be chain reacting purely on a basis of calculation. The practical feasibility of this program did not appear too hopeful. We know now that the positive excess that makes possible a chain reaction in a graphite-uranium system is of only a few percent. Since many factors of absorption and production of neutrons enter in the final result, it is clear that they should be known individually with extreme accuracy to make a prediction possible. The measuring methods developed up to 1941 seldom permitted the measurement of nuclear properties with an accuracy better than 10 percent and were therefore inadequate to give a basis for calculations that would permit answering in a reliable way the question of whether the chain reaction with natural uranium and graphite was or was not possible.

In any system of finite dimensions some neutrons escape by diffusing out of its surface. The loss of neutrons by escape can in principle be eliminated by increasing the size of the system. It was clear in 1941 that the balance of neutrons capable of sustaining a chain reaction, even if at all positive, would be so small as to make it necessary to use a system of very large size in order to eliminate most of the loss of neutrons by escape. It was important to devise methods capable of answering the following questions: (1) whether a system containing lumps of uranium distributed through the graphite in a given lattice arrangement would become chain reacting provided its dimensions were infinitely large, and (2) assuming

First Chain-Reacting Pile

a positive answer to the previous question, what minimum dimensions would be needed actually to achieve the chain reaction? The minimum dimensions are usually called the critical size of the pile. Since the method of detailed calculation from the values measured for the constants was inadequate as explained before, one had to devise some way that would give more directly the required answers.

A brute-force method for this would be to set up a system of the given structure and keep on adding to it until a chain reaction actually is achieved or the system refused to react even when built up to enormous size. This method obviously would be exceedingly expensive both in materials and labor. Fortunately it is possible to obtain a fairly accurate answer to the two questions by using a relatively small sample of the structure under investigation. The first experiments of this type, the so-called intermediate or exponential experiments, were set up at Columbia University in the summer and fall of 1941. A lattice structure was set up containing cans filled with uranium oxide spread throughout a mass of some thirty tons of graphite. A primary source of neutrons was inserted at the bottom of this mass and the distribution of the neutrons throughout the mass was investigated in detail and compared with the theoretical expectation.

The results of this first experiment was somewhat discouraging in that it indicated that a system of that structure, even if built up to infinite size, would still have a negative balance of neutrons and more precisely a loss of 13 percent of the neutrons each generation. In spite of this negative result hope was not abandoned. Indeed sizable improvements to this first structure could be expected as indicated below.

Early in 1942 all the groups working on the production of a chain reaction were united at the Metallurgical Laboratory of the University of Chicago under the general leadership of Arthur Compton. During 1942 some twenty or thirty exponential experiments were carried out at Chicago in the attempt to improve on the conditions of the first experiment. Two different types of improvements were pursued. One consisted in a better adjustment of the dimensions of the lattice and the other in the use of better materials. Impurities had to be eliminated to a surprisingly high extent from both uranium and graphite since the parasitic absorption due to elements appearing as common impurities in uranium and graphite was responsible for a loss of an appreciable fraction of the neutrons. The problem was tackled to organize large-scale production of many tons of graphite and uranium of an unprecedented purity. Also the production of uranium in metallic form was vigorously pursued. Up to 1941

First Chain-Reacting Pile

uranium metal had been produced only in very small amounts, often of questionable purity. Uranium metal was mostly produced in the form of a highly pyrophoric powder which in several cases burst spontaneously into flames when coming in contact with air. These pyrophoric properties were only somewhat reduced by sintering the powder into compact blocks. Some of these sintered blocks were used in exponential experiments carried out in order to obtain information on the properties of a system containing metallic uranium; while the experiments were in progress the blocks were burning so fast that they felt hot to the touch and we were afraid that they might actually burst into flames before we could go through with the experiment.

Toward the fall of 1942 the situation as to the production of materials gradually improved. Through the joint efforts of the staff of the Metallurgical Laboratory and of several industrial firms, better and better graphite was obtained. Industrial production of practically pure uranium oxide was organized and some amount of cast uranium metal was produced. The results of the exponential experiments improved correspondingly to the point that the indications were that a chain reacting unit could be built using these better brands of materials.

The actual erection of the first chain reacting unit was initiated in October 1942. It was planned to build a lattice structure in the form of a huge sphere supported by a wooden structure. The structure was to be erected in a Squash Court on the campus of the University of Chicago. Since we were somewhat doubtful whether the dimensions as planned would be sufficiently large, the structure was actually built inside a huge tent of balloon cloth fabric that in case of need could have been sealed for the purpose of removing the air in order to avoid the parasitic absorption of the atmospheric nitrogen. This precaution actually proved unnecessary.

It took a little over one month to build the structure. A large number of physicists, among them W. H. Zinn, H. L. Anderson, and V. C. Wilson, collaborated in the construction. During this time the approach to the chain reacting conditions was followed day by day by measuring the neutron intensity building up inside the pile. Some neutrons are produced spontaneously by uranium in very small numbers. When the system approaches the critical size, each of these neutrons multiplies for several generations before final absorption. Indeed, when the reproduction factor of the pile is, for instance, 99 percent, each neutron multiplies in the average one hundred generations. Consequently, the density of neutrons increases throughout the mass as the critical dimensions are approached and

First Chain-Reacting Pile

tends to diverge at the critical size. By watching the rise of the neutron density, one obtains, therefore, a positive method for extrapolating to the critical size.

Appreciably before the dimensions originally planned for the structure were reached, the measurements of the neutron density inside the structure indicated that the critical size would soon be attained. From this time on work was continued under careful supervision so as to make sure that criticality would not be inadvertently reached without proper precautions. Long cadmium strips were inserted in slots that had been left for this purpose in the structure. Cadmium is one of the most powerful absorbers of neutrons and the absorption of these strips was large enough to make sure that no chain reaction could take place while they were inside the pile. Each morning the cadmium strips were slowly removed, one by one, and a determination of the neutron intensity was carried out in order to estimate how far we were from the critical conditions.

On the morning of December 2, 1942, the indications were that the critical dimensions had been slightly exceeded and that the system did not chain react only because of the absorption of the cadmium strips. During the morning all the cadmium strips but one were carefully removed; then this last strip was gradually extracted, close watch being kept on the intensity. From the measurements it was expected that the system would become critical by removing a length of about eight feet of this last strip. Actually when about seven feet were removed the intensity rose to a very high value but still stabilized after a few minutes at a finite level. It was with some trepidation that the order was given to remove one more foot and a half of the strip. This operation would bring us over the top. When the foot and a half was pulled out, the intensity started rising slowly, but at an increasing rate, and kept on increasing until it was evident that it would actually diverge. Then the cadmium strips were again inserted into the structure and the intensity rapidly dropped to an insignificant level.

This prototype of a chain reacting unit proved to be exceedingly easy to control. Intensity of its operation could be adjusted with extreme accuracy to any desired level. All the operator has to do is to watch an instrument that indicates the intensity of the reaction and move the cadmium strips in if the intensity shows a tendency to rise, and out if the intensity shows a tendency to drop. To operate a pile is just as easy as to keep a car running on a straight road by adjusting the steering wheel when the car tends to shift right or left. After a few hours of practice an operator can keep easily the intensity of the reaction constant to a very small fraction of 1 percent.

The first pile had no device built in to remove the heat produced

First Chain-Reacting Pile

by the reaction and it was not provided with any shield to absorb the radiations produced by the fission process. For these reasons it could be operated only at a nominal power which never exceeded two hundred watts. It proved, however, two points: that the chain reaction with graphite and natural uranium was possible, and that it was very easily controllable.

A huge scientific and engineering development was still needed to reduce to industrial practice the new art. Through the collaboration of all the men of the Metallurgical project and of the du Pont Company, only about two years after the experimental operation of the first pile large plants based essentially on the same principle were put in operation by the du Pont Company at Hanford, producing huge amounts of energy and relatively large amounts of the new element, plutonium.

Notes

1. FAMILY BACKGROUND AND YOUTH

1. I am indebted to the Italian State Railroads (Ferrovie dello Stato) for personal information on Alberto Fermi. See also Laura Fermi, *Atoms in the Family* (Chicago: University of Chicago Press, 1954), chap. 2; and private letters of Maria Fermi.

2. Enrico Persico, "Souvenir de Enrico Fermi" *Scientia* 90 (1955): 316.

3. Theodor Reye, *Geometria di Posizione,* trans. A. Faifofer (Venice, 1884).

4. Compare the postcard dated August 18, 1918, in Appendix 1.

5. Part of this essay is reproduced photographically in *The Collected Papers of Enrico Fermi,* 2 vols. (Chicago: University of Chicago Press, 1962, 1965) vol. 1, facing p. xxii.
This two volume work contains Fermi's writings, chronologically ordered and numbered. In referring to a specific paper in this collection we will use the abbreviated form *FP* 123, where the number is that of the cited paper. When referring to a particular page in the collection we will use the form *FP*, 1:78, where the volume is separated from the page reference by a colon.

6. Laura Fermi, *Atoms in the Family,* p. 24. See also Thelma Nason, "A Man for All Sciences" The Johns Hopkins Magazine, v.*17*, Number 4, p. 12 (1966).

7. *FP,* 1:55–56.

2. APPRENTICESHIP

1. Translation from *FP* 120.

2. I am indebted to Professor Epicarmo Corbino for some of the information on O. M. Corbino's youth. See also Orso Mario Corbino, *Conferenze e Discorsi* (Rome: Enzo Pinci, 1937).

3. For an obituary of D. Macaluso see Luciano Sesta, "In Memoria di D. Macaluso," *Nuovo Cimento* 10 (1933): 1.

4. For an obituary of Righi by Corbino see *Rendiconti Lincei* 30 (1921): 215.

Notes: Chapter 3

5. Corbino, *Conferenze e Discorsi*, p. 167.

6. Laura Fermi, *Atoms in the Family*, p. 30.

7. Ministero della Pubblica Istruzione, *Bollettino Ufficiale, Atti di Amministrazione*, Anno 50 (March 8, 1923) 1:802.

8. "Proof that a Normal Mechanical System is in General Quasi-ergodic" (*FP* 11).

9. "On the Theory of Collisions between Atoms and Charged Particles" (*FP* 23a).

10. See letter dated July 26, 1924, in Appendix 1.

11. Niels Bohr, *Zeitschrift für Physik* 34 (1925): 149.

12. "Effect of an Alternating Magnetic Field on the Polarization of the Resonance Radiation of Mercury Vapor" (*FP* 26).

13. *FP*, 1:159.

14. Ministero della Pubblica Istruzione, *Bollettino Ufficiale, Atti di Amministrazione*, Anno 53 (March 4, 1926) 1:793.

15. See letters dated June 1925 and later in Appendix 1.

16. Erwin Schrödinger, "Quantisierung als Eigenwert Problem," *Annalen der Physik*, 4th ser., 79 (1926): 361. The first paper is dated January 27, 1926. See also Erwin Schrödinger, *Abhandlungen zur Wellenmechanik* (Leipzig: J. A. Barth Verlag, 1928).

17. Karl Przibram, *Briefe zur Wellenmechanik* (Wien:Springer, 1963). For a history of quantum mechanics, see Max Jammer, *The Conceptual Development of Quantum Mechanics* (New York: McGraw-Hill, 1966).

18. "On the Wave Mechanics of Collisions" (*FP* 36).

19. Ministero della Pubblica Istruzione, *Bollettino Ufficiale, Atti di Amministrazione*, Anno 54 (March 3, 1927) 1:634.

3. Professor at Rome

1. Paul Drude, *Lehrbuch der Optik* (Leipzig: S. Hirzel, 1900).

2. J. J. Thomson, *Conduction of Electricity through Gases* (London: Cambridge University Press, 1903).

3. Fritz Reiche, *Die Quantentheorie* (Berlin: Springer, 1921).

4. See *Atti del Congresso Internazionale dei fisici 1927* (Bologna: Zanichelli, 1928), especially p. 470–71.

5. See also E. Amaldi, *La vita e l'opera di Ettore Majorana* (Rome: Accademia Nazionale dei Lincei, 1966).

6. "A Measurement of the h/k Ratio through Anomalous Dispersion of Thallium" (*FP* 40).

7. See Maurice de Broglie, *Les premiers Congrès de Physique Solvay* (Paris: Albin Michel, 1951).

8. *Proceedings of the Royal Society* (London), ser. A, 114 (1927): 243, 710.

Notes: Chapter 3

9. H. A. Bethe, "Memorial Symposium in Honor of E. Fermi at the Washington Meeting of the American Physical Society, April 29, 1955," *Reviews of Modern Physics* 27 (1955): 253.

10. E. P. Wigner, *Yearbook of the American Philosophical Society* (Philadelphia: American Philosophical Society, 1955), pp. 435–39.

11. H. A. Bethe, private communication.

12. See E. Amaldi, "Georges Placzek," *Ricerca Scientifica* 26 (1956): 2038.

13. See "The Adiabatic Principle and Kinetic Energy in the New Wave Mechanics" (*FP* 37); "Quantum Mechanics and the Magnetic Moment of Atoms" (*FP* 39); and "On the Mechanism of Emission in Wave Mechanics" (*FP* 42).

14. See, for example, Max Jammer, *The Conceptual Development of Quantum Mechanics* (New York: McGraw-Hill, 1966).

15. O. M. Corbino, "I nuovi compiti della fisica sperimentale," *Atti Società Italiana Progresso delle Scienze* 18 (1929): 1,157; also *L'Energia Elettrica*, October 1929, p. 998.

16. In 1929 these were the accepted ideas. The neutron was discovered only in 1932.

17. Antonio Garbasso, "Poche parole di un fisico agli elettrotecnici," *L'Elettrotecnica* 16 (1929): 717.

18. O. M. Corbino, "A proposito di due discorsi sulla situazione della Fisica," *L'Elettrotecnica* 16 (1929): 772.

19. *Reale Accademia d'Italia, Fondazione A. Volta Convegno di Fisica Nucleare Roma 1932* (Rome: Accademia d'Italia, 1932).

20. "The Present State of Nuclear Physics" (*FP* 72b).

21. *FP*, 1:488.

22. Ibid.

23. See, for example, C. S. Wu and S. A. Moszkowski, *Beta Decay* (New York: Interscience, 1966), p. 385.

24. Hideki Yukawa, "On the Interaction of Elementary Particles," *Progress of Theoretical Physics* 17 (1935): 48; see also H. Yukawa and K. Chihiro, "The Birth of the Meson Theory," *American Journal of Physics* 18 (1950): 154.

25. Wigner, *Yearbook of the American Philosophical Society*, pp. 435–39.

26. Irène Joliot-Curie and Frédéric Joliot-Curie, "Un nouveau type de radioactivité," *Comptes Rendus* 198 (1934): 254; and "Artificial Production of a New Type of Radioelements," *Nature*, 133 (1934): 201.

27. "Radioactivity Produced by Neutron Bombardment. I" (*FP* 84–94).

28. Years later Maria Mayer, at Fermi's instigation, calculated the energy of the $5f$ orbits by the Fermi-Thomas statistical method, as Fermi had done (*FP* 47) for the $4f$ orbits of the rare earths. She thus gave a theoretical foundation to the semi-empirical arguments of Abelson, McMillan, and others. See M. Mayer, *Physical Review* 60 (1941): 184.

29. Ida Noddack, *Angewandte Chemie* 47 (1934): 653.

Notes: Chapter 4

30. Corbino, *Conferenze e Discorsi*, p. 51.

31. Laura Fermi, *Atoms in the Family*, p. 92.

32. "Artificial Radioactivity Produced by Neutron Bombardment" (*FP* 98).

33. *FP*, 2:927.

34. Papers were dated in Fascist style: XII means twelfth year of the Fascist era.

35. Lewis L. Strauss, *Men and Decisions* (Garden City, N.Y.: Doubleday, 1962), p. 165.

36. "Artificial Radioactivity Produced by Neutron Bombardment. II" (*FP* 107).

37. "On the Pressure Shift of Lines High in Spectral Series" (*FP* 95).

38. *FP*, 1:808.

39. "The Nucleus" (*FP* 247).

40. *FP*, 1:810.

41. Piotr L. Kapitza, "Recollections of Lord Rutherford," *Nature* (London), 210 (1966): 780. Rutherford's letters are at p. 782.

42. E. Fermi, *Thermodynamics* (New York: Prentice Hall, 1937).

43. "Un Maestro: Orso Mario Corbino" (*FP*, 120).

44. As background information see Denis Mack Smith, *Italy: A Modern History* (Ann Arbor, Mich.: University of Michigan Press, 1959).

45. Translation by Muriel Kittel.

46. For background information see Renzo DeFelice, *Storia degli Ebrei Italiani sotto il Fascismo* (Turin: G. Einaudi, 1961).

47. Otto Hahn and Fritz Strassmann, "Über den Nachweis und das Verhalten der bei der Bestrahlung des Urans mittels Neutronen entstehenden Erdalkalimetallen," *Naturwissenschaften* 27 (1939): 711.

48. See L. A. Turner, "Nuclear fission," *Reviews of Modern Physics* 12 (1940): 1; O. Frisch, "The Interest Is Focusing on the Atomic Nucleus," in *Niels Bohr,* ed. S. Rozental (New York: John Wiley & Sons, 1967), p. 137; and Otto Hahn, *Vom Radiothor zur Uranspaltung: Eine wissenschaftliche Selbstbiographie* (Braunschweig: V. Vieweg, 1962).

49. Laura Fermi, *Atoms in the Family*, p. 139.

4. Emigration and the War Years

1. S. K. Allison, "Enrico Fermi 1901–1954," *National Academy of Sciences, United States of America, Biographical Memoirs* 30 (1957): 125.

2. Strauss, *Men and Decisions*, p. 240.

3. See *FP* 269.

4. *FP*, 2 :1.

5. See also Leo Szilard, "Reminiscences," *Perspectives in American History* 2 (1968): 94.

Notes: Chapter 4

6. See S. Ulam, "John von Neumann," *Bulletin of the American Mathematical Society* 64 (1958): 1; and S. Ulam, H. W. Kuhn, A. W. Tucker, and C. E. Shannon, "John von Neumann 1903–1957," *Perspectives in American History,* 2 (1968): 235.

7. *FP,* 2:11.

8. Fermi, in a letter to A. H. Compton dated August 31, 1943, says so.

9. Some of the patents are: Fermi-Szilard, neutronic reactor, pat. 2,708,656; Fermi-Anderson, testing materials in neutronic reactor, pat. 2,768,134; Fermi, exponential pile, pat. 2,780,595; Fermi-Zinn, neutronic reactor shield, pat. 2,807,727; Fermi-Leverett, chain reacting system, pat. 2,837,477; Fermi-Szilard, air-cooled neutron reactor, pat. 2,836,554; Fermi-Leverett, method of sustaining a neutronic chain reacting system, pat. 2,813,070; Fermi-Szilard, neutronic reactor, pat. 2,807,581; Fermi-Zinn, neutronic reactor, pat. 2,852,461; Fermi, neutronic reactor, pat. 2,931,762.

10. See for example, Arthur H. Compton, *Atomic Quest* (New York: Oxford University Press, 1956), p. 36.

11. Compton, *Atomic Quest,* p. 28.

12. The history of the United States Atomic Energy Project up to 1945 is admirably told in Henry De Wolf Smyth, *Atomic Energy for Military Purposes* (Princeton, N. J.: Princeton University Press, 1945); see also R. G. Hewlett and E. O. Anderson, Jr., *The New World 1939–1946* (University Park, Pa.: Pennsylvania State University Press, 1962); G. Hewlett and F. Duncan, *Atomic Shield, 1946–51* (Pennsylvania State University Press, 1969). Also Margaret Gowing, *Britain and Atomic Energy* (London: St. Martin's Press, 1964); David Irving, *The Virus House* (London: William Kiber, 1967).

13. Strauss, *Men and Decisions,* p. 172.

14. "Neutron Production and Absorption in Uranium" (*FP* 132). See also "Production of Neutrons in Uranium Bombarded by Neutrons" (*FP* 130) and "Simple Capture of Neutrons by Uranium" (*FP* 131).

15. *FP,* 2:11.

16. Hewlett and Anderson, *The New World,* p. 17.

17. See note 12 above.

18. For a short biography of E. O. Lawrence, see L. W. Alvarez, *National Academy of Sciences, United States of America, Biographical Memoirs,* in press.

19. "Fission of Uranium by Alpha Particles" (*FP* 135).

20. In *FP,* 1 : XXVII and *FP,* 2 : 1081 the year 1939 should read 1940.

21. Smyth, *Atomic Energy for Military Purposes,* par. 4.24.

22. Smyth, *Atomic Energy for Military Purposes,* par 4.25.

23. Ibid, par. 4.51.

24. Winston S. Churchill, *The Second World War,* vol. 4, *The Hinge of Fate* (London: Cassel, 1951), p. 3.

25. Smyth, *Atomic Energy for Military Purposes,* par. 5.19.

Notes: Chapter 4

26. Ibid, par. 5.21.

27. See Leslie R. Groves, *Now It Can Be Told* (New York: Harper & Brothers, 1962).

28. "The Development of the First Chain Reacting Pile" (*FP* 223), reproduced in its entirety in Appendix 4.

29. *FP*, 2 : 268.

30. *FP* 160–80.

31. For a list of those present see *FP*, 2: 269.

32. In a speech in Chicago, December 2, 1967, commemorating the twenty-fifth anniversary of the pile.

33. Compton, *Atomic Quest*, p. 144.

34. "Production of Low Energy Neutrons by Filtering through Graphite" (*FP* 191).

35. "A Thermal Neutron Velocity Selector and Its Application to the Measurement of the Cross Section of Boron" (*FP* 200).

36. "Reflection of Neutrons on Mirrors" (*FP* 220).

37. Hewlett and Anderson, *The New World*, p. 204.

38. An informative autobiography of Oppenheimer is given in *In the Matter of J. R. Oppenheimer, Transcript of a Hearing before Personnel Security Board, Washington, April 24, 1954, through May 6, 1954* (Washington, D.C.: Government Printing Office, 1954), p. 26–57. See also R. Serber, V. F. Weisskopf, A. Pais, and G. T. Seaborg, *Oppenheimer* (New York: Charles Scribner's Sons, 1969) and Denise Royal, *The Story of Robert Oppenheimer* (New York: St. Martin 1969).

39. R. Serber, "The Los Alamos Primer," Los Alamos unpublished report, now declassified.

40. "A Course on Neutron Physics" (*FP* 222).

41. Aage Bohr, "The War Years," in *Niels Bohr*, ed. S. Rozental (Amsterdam: North Holland Publishing Co., 1967), p. 191.

42. H. A. Compton and S. K. Allison, *X-rays in Theory and Experiment* (New York: Van Nostrand, 1935). Allison once jokingly said to Fermi that his dissertation, which as we have seen was on X-rays, could not have been much good since it was not quoted in Compton and Allison, to which Fermi immediately replied that this showed merely an inadequacy of that volume.

43. Hewlett and Anderson, *The New World*, p. 251.

44. Groves, *Now It Can Be Told*, p. 293.

45. Ibid., p. 414.

46. A. Bohr, "The War Years," p. 191.

47. Alice Kimball Smith, *A Peril and a Hope* (Chicago and London: University of Chicago Press, 1965), p. 539.

48. Hewlett and Anderson, *The New World*, ch. 10.

49. Churchill, *The Hinge of Fate*, p. 561.

50. Smith, *A Peril and a Hope*, p. 34.

Notes: Chapter 5

51. Ibid, p. 36.
52. Hewlett and Anderson, *The New World*, p. 356–60.
53. Smith, *A Peril and a Hope*, p. 560.
54. Ibid, p. 49.

5. Professor at Chicago

1. For an early description of the institutes see Samuel K. Allison, "Institute for Nuclear Studies, The University of Chicago," *Scientific Monthly* 65 (1947): 482.
2. Samuel K. Allison, "Enrico Fermi, 1901–1954," *Physics Today* 8 (1955): 9.
3. Smith, *A Peril and a Hope*, p. 89.
4. Hewlett and Anderson, *The New World*, p. 422.
5. Smith, *A Peril and a Hope*, p. 96.
6. Smyth, *Atomic Energy for Military Purposes*, par. 13.8.
7. Smith, *A Peril and a Hope*, p. 142.
8. Ibid., p. 383.
9. *In the Matter of J. R. Oppenheimer*, p. 79–80.
10. See for instance J. Stefan Dupré and Sanford A. Lakoff, *Science and the Nation: Policy and Politics* (Englewood Cliffs, N. J.: Prentice-Hall, 1962), ch. 8.
11. Stan M. Ulam, "Thermonuclear Devices," in *Perspectives in Modern Physics*, ed. R. E. Marshak (New York: Interscience, 1966), p. 593ff.
12. E. Fermi, *Notes on Quantum Mechanics* (Chicago: University of Chicago Press, 1961).
13. A. J. Cronin, D. F. Greenberg, and V. L. Telegdi, *University of Chicago Graduate Problems in Physics: With Answers* (Reading, Mass.: Addison Wesley, 1967).
14. *FP*, 2:673. I had similar experiences when I substituted for Fermi in some courses. Furthermore, the subjects emphasized by Fermi in his private Chicago seminars considerably overlap those he treated in his private seminars in Rome. They covered what he considered the most important topics for a future research physicist.
15. "Phase of Neutron Scattering" (*FP* 227); "Interference Phenomena of Slow Neutrons" (*FP* 228); "Phase of Scattering of Thermal Neutrons by Aluminum and Strontium" (*FP* 229); "Spin Dependence on Scattering of Slow Neutrons by Be, Al, and Bi" (*FP* 230); "On the Interaction between Neutrons and Electrons" (*FP* 234).
16. Maria Goeppert Mayer, "The Shell Model," in *Les Prix Nobel en 1963* (Stockholm: Imprimerie Royale Norstedt and Söner, 1964), p. 133ff.
17. "The Decay of Negative Mesotrons in Matter" (*FP* 232).
18. "The Capture of Negative Mesotrons in Matter" (*FP* 233).

Notes: Appendix 1

19. "The Origin of Cosmic Radiation" (*FP* 237).

20. E. P. Wigner, *Symmetries and Reflections* (Bloomington, Ind.: Indiana University Press, 1967), p. 171.

21. "Conference on Atomic Physics" (*FP* 240).

22. E. Fermi, *Elementary Particles* (New Haven, Conn.: Yale University Press, 1951).

23. "High Energy Nuclear Events" (*FP* 241).

24. "Total Cross Sections of Positive Pions in Hydrogen" (*FP* 250).

25. See Dupré and Lakoff, *Science and the Nation,* chap. 7, for background information.

26. *Physics Today,* May 1953, pp. 20–26; ibid., June 1953, p. 20. Samuel A. Laurence, *The Inter-University Case Program No. 68: The Battery Additive Controversy* (University, Ala.: University of Alabama Press, 1962) gives a detailed history of the affair.

27. *In the Matter of J. R. Oppenheimer, Transcript of a Hearing before Personnel Security Board, Washington, April 12 through May 6, 1954* (Washington, D.C. Government Printing Office, 1954). Fermi's testimony is on pp. 394–98. See also, for background information, J. S. Dupré and S. A. Lakoff, *Science and the Nation.*

28. "Galactic Magnetic Fields and the Origin of Cosmic Radiation," (*FP* 265); "Multiple Production of Pions in Pion-Nucleon Collisions" (*FP* 263); "Studies of Nonlinear Problems" (*FP* 266).

29. "Lectures on Pions and Nucleons" (*FP* 270).

30. On learning of Fermi's illness Commissioner Strauss recommended to President Eisenhower an award to Fermi, in conformity with a provision of the Atomic Energy Act. The President acted promptly and succeeded in giving the award of $25,000 to Fermi before his death. Later the award, supplemented by a medal, was called the Fermi award and was conferred annually.

31. A version consisting of notes compiled by J. O'Rear, A. H. Rosenfeld, and R. A. Schluter (three of Fermi's students) and not revised by Fermi is E. Fermi, *Nuclear Physics* (Chicago: University of Chicago Press, 1949).

Appendix 1

1. Professor Eredia had been Fermi's physics professor at the liceo. He then became director of the metereological office, located near the Vittorio Emanuele Library, one of the great State libraries of Italy. It is probable that Professor Eredia had recognized Fermi's talent and was happy to help him.

2. The barometer was similar to a normal mercury barometer except that it used water instead of mercury and the upper, evacuated space was replaced by a vessel containing some air saturated with water vapor. Thus it could be used only if one carefully measured both the height of the water and the temperature. The calibration was far from simple. The boys had developed the formula for deriving the pressure from the temperature and the height of the water column.

3. This was for high school tests.

4. Chwolson's physics treatise.

Notes: Appendix 1

5. A dialect phrase: "we will think about it."

6. The reference to the earth's magnetic field is to a measurement done at home by the magnetometer method. They had found that $h = 0.2216$ gauss, with an estimated error of ± 0.03; the correct value was 0.23.

7. P. Appell, *Traité de mécanique rationelle* (Paris: Gauthier-Villars, 1909).

8. S. D. Poisson, *Traité de mécanique,* 1st ed. (Paris, 1811).

9. Persico had posed the problem of hitting a ball in an arbitrary position by bouncing another ball on the rim of a circular billiard table.

10. See n. 8 above.

11. This is an illustration of stroboscopic methods. There is no mention of the application to the determination of surface tension.

12. See n. 7 above.

13. This may refer to his letter of February 12, 1919.

14. The original was written in German and contained several language errors.

15. The granting of the doctor's degree to Persico.

16. The illustration shows two hands clasped in a congratulatory handshake.

17. There was a joke about a navy captain who, as his ship entered a mine field, gave the command: "Forward, *slowly,* almost backwards."

18. Fermi himself.

19. "Who cares?"

20. The Alto Adige or South Tyrol.

21. Persico had thought of studying some properties of ionized gases by using grids of parallel wires at potentials V_1 and V_2 alternately.

22. *Società antiprossimo:* antineighbor "society for the Harassment of Neighbors."

23. Anti prossimo: Antineighbor.

24. A student.

25. *Physikalische Zeitschrift* 23 (1922): 340.

26. See *FP* 6 and 7.

27. *FP* 38*b*.

28. The treatise is by Houzeau, and the problem had not been previously treated.

29. This was the separate dissertation for the Scuola Normale mentioned in the preceding letter.

30. An astronomy professor at the University of Pisa.

31. F. Tisserand, *Traité de Mécanique celeste* (Paris: Gauthier-Villars, 1891).

Notes: Appendix 1

32. A caricature of Don Sturzo, founder of the Christian-Democratic party.

33. Allusion to an Italian proverb: "Whoever keeps silent agrees."

34. *Pastore,* of course, means both "pastor" and "shepherd."

35. A proposed bicycle tour in Umbria with Rasetti and Persico.

36. The competition was for a fellowship abroad, which Fermi won (see page 32).

37. Fermi's sister.

38. A friend and schoolmate at Pisa, later professor of physics in Florence.

39. Fermi's mother had recently died, and he and his sister were moving to another part of Rome. Professor Trabacchi was a physicist at the "Istituto di Sanità Pubblica." (see p. 33.)

40. The Ministry of Public Instruction in Rome.

41. The expenses of the committee awarding the *libera docenza* were charged to the candidates and were to be paid before the award was made.

42. A physicist in the famous Leyden Cryogenic Laboratory.

43. Professor Giovanni Gentile was Minister of Education under Mussolini. He sponsored a vast reform in the Italian educational system.

44. A professor of mathematics and a friend of Fermi.

45. *FP* 28.

46. Classical Maturity (*maturità classica*). This newly established comprehensive examination tested all students graduating from high school. Passing this examination was a prerequisite for admission to a university.

47. The prevailing rate of exchange: 147 Italian lire per pound sterling.

48. That is, in English. Italians have the impression that English is spoken with a contracted mouth or with a potato in the mouth. The sounds of English are deemed inhuman, thus the phrase.

49. Levi-Civita's sister-in-law.

50. The high school Fermi and his sister had attended.

51. President of the Catholic University in Milan.

52. Professor of ophthalmology, influential in academic politics.

53. The Minister of Education.

54. It was Ornstein's, a physicist at Utrecht.

55. Professor of experimental physics at Florence and mayor of that city.

56. Professors of experimental physics at Naples and Bologna, respectively.

57. Professor of mathematical physics at Turin.

58. Professor of theoretical mechanics at Pisa.

59. Primarily a mathematical physicist.

60. From Naples, where he later became professor of experimental physics.

Notes: Appendix 1

61. The Sella prize was a small annual prize for young Italian physicists, administered by the Accademia dei Lincei.

62. The idea was to concentrate sunlight, using a war-surplus light projector, on a discharge tube and to observe the Stark effect produced by the electric field of sunlight.

63. The appointment was in fact irregular and could have nullified the competition. After some hesitation, Fermi and Persico pointed out this danger to Levi-Civita, who was then replaced by Maggi.

Bibliography

ARTICLES, LECTURES, AND OTHER SHORTER WORKS

The following titles are numbered as they appear in the Contents and Bibliography of *The Collected Works of Enrico Fermi*. Fermi published only those titles for which a source is given. The titles for which no sources are cited remained unpublished in his lifetime, either because they were classified or because, being merely reports of work in progress, they were not considered by Fermi to deserve publication—or for some other reason.

1. "Sulla dinamica di un sistema rigido di cariche elettriche in moto traslatorio." *Nuovo Cimento* 22 (1921): 199–207. *FP*, 1:1–7.
2. "Sull'elettrostatica di un campo gravitazionale uniforme e sul peso delle masse elettromagnetiche." *Nuovo Cimento* 22 (1921): 176–88. *FP*, 1:8–16.
3. "Sopra i fenomeni che avvengono in vicinanza di una linea oraria." *Rend. Lincei* 31 (1) (1922): 21–23, 51–52, 101–3. *FP*, 1:17–23.
4a. "Über einen Widerspruch zwischen der elektrodynamischen und der relativistischen Theorie der elektromagnetischen Masse." *Phys. Zeits.* 23 (1922): 340–44.
4b. "I. Correzione di una grave discrepanza tra la teoria delle masse elettromagnetiche e la teoria della relatività. Inerzia e peso della elettricità." *Rend. Lincei* 31 (1) (1922): 184–87; "II. Correzione di una grave discrepanza tra la teoria elettrodinami ca e quella relativistica delle masse elettromagnetiche. Inerzia e peso dell'elettricità." *Rend. Lincei* 31 (1) (1922): 306–9.
4c. "Correzione di una contraddizione tra la teoria elettrodinamica e quella relativistica delle masse elettromagnetiche." *Nuovo Cimento* 25 (1923): 159–70. *FP*, 1:24–32.
5. "Le masse nella teoria della relatività." Appendix to Italian edition of *I fondamenti della relatività Einsteiniana*, by A. Kopff, pp. 342–44. Milan: Hoepli, 1923. *FP*, 1:33–34.
6. "I raggi Röntgen." *Nuovo Cimento* 24 (1922): 133–63. *FP*, 1:35–54.

Bibliography

7. "Formazione di immagini coi raggi Röntgen." *Nuovo Cimento* 25 (1923): 63–68. *FP*, 1:55–59.
8. "Sul peso dei corpi elastici." *Memorie Lincei* 14 (1923): 114–24. *FP*, 1:60–71.
9. "Sul trascinamento del piano di polarizzazione da parte di un mezzo rotante." *Rend. Lincei* 32 (1) (1923): 115–18. *FP*, 1:72–75.
10. "Sulla massa della radiazione in uno spazio vuoto " With A. Pontremoli. *Rend. Lincei* 32 (1) (1923). 162–64. *FP*, 1:76–78.
11a. "I. Beweis dass ein mechanisches Normalsystem im allgemeinen quasi-ergodisch ist." *Phys. Zeits.* 24 (1923): 261–65; "Über die Existenz quasi-ergodischer Systeme." *Phys. Zeits.* 25 (1924): 166–67. *FP*, 1:79–87.
11b. "Dimostrazione che in generale un sistema meccanico normale è quasi ergodico." *Nuovo Cimento* 25 (1923): 267–69.
12. "Il principio delle adiabatiche ed i sistemi che non ammettono coordinate angolari." *Nuovo Cimento* 25 (1923): 171–75. *FP*, 1:88–91.
13. "Alcuni teoremi di meccanica analitica importanti per la teoria dei quanti." *Nuovo Cimento* 25 (1923): 271–85. *FP*, 1:92–101.
14. "Sulla teoria statistica di Richardson dell'effetto fotoelettrico." *Nuovo Cimento* 26 (1923): 97–104. *FP*, 1:102–7.
15. "Generalizzazione del teorema di Poincaré sopra la non esistenza di integrali uniformi di un sistema di equazioni canoniche normali." *Nuovo Cimento* 26 (1923): 105–15. *FP*, 1:108–13.
16. "Sopra la teoria di Stern della costante assoluta dell'entropia di un gas perfetto monoatomico." *Rend. Lincei* 32 (2) (1923): 395–98. *FP*, 1:114–17.
17a. "Sulla probabilità degli stati quantici." *Rend. Lincei* 32 (2) (1923): 493–95. *FP*, 1:118–20.
17b. "Über die Wahrscheinlichkeit der Quantenzustände." *Z. Physik* 26 (1924): 54–56.
18. "Sopra la riflessione e la diffusione di risonanza." *Rend. Lincei* 33 (1) (1924): 90–93. *FP*, 1:121–23.
19. "Considerazioni sulla quantizzazione dei sistemi che contengono degli elementi identici." *Nuovo Cimento* 1 (1924): 145–52. *FP*, 1:124–29.
20. "Sull'equilibrio termico di ionizzazione." *Nuovo Cimento* 1 (1924): 153–58. *FP*, 1:130–33.
21a. "Berekeningen over de intensiteiten van spektraallijnen." *Physica* 4 (1924): 340–43.
21b. "Sopra l'intensità delle righe multiple." *Rend. Lincei* 1 (1925): 120–24. *FP*, 1:134–37.
22. "Sui principi della teoria dei quanti." *Rend. Seminario matematico Università di Roma* 8 (1925): 7–12. *FP*, 1:138–41.
23a. "Sulla teoria dell'urto tra atomi e corpuscoli elettrici." *Nuovo Cimento* 2 (1925): 143–58, also *Rend. Lincei* 1 (1924): 243–45.
23b. "Über die Theorie des Stosses zwischen Atomen und elektrisch geladenen Teilchen." *Z. Physik* 29 (1924): 315–27. *FP*, 1:142–53.

Bibliography

24. "Sopra l'urto tra atomi e nuclei di idrogeno." *Rend. Lincei* 1 (1925): 77–80. *FP*, 1:154–56.
25. "Una relazione tra le costanti delle bande infrarosse delle molecole triatomiche." *Rend. Lincei* 1 (1925): 386–87. *FP*, 1:157–58.
26. "Effect of an Alternating Magnetic Field on the Polarization of the Resonance Radiation of Mercury Vapour." With F. Rasetti. *Nature* (London) 115 (1925): 764 (letter). *FP*, 1:159–60.
27. "Über den Einfluss eines wechselnden magnetischen Feldes auf die Polarisation der Resonanzstrahlung." With F. Rasetti. *Z. Physik* 33 (1925): 246–50.
28(1). "Effetto di un campo magnetico alternato sopra la polarizzazione della luce di risonanza." With F. Rasetti. *Rend. Lincei* 1 (1925): 716–22. *FP*, 1:161–66.
28(2). "Ancora dell'effetto di un campo magnetico alternato sopra la polarizzazione della luce di risonanza." With F. Rasetti. *Rend Lincei* 2 (1925): 117–20. *FP*, 1:167–70.
29. "Sopra la teoria dei corpi solidi." *Periodico di Matematiche* 5 (1925): 264–74. *FP*, 1:171–77.
30. "Sulla quantizzazione del gas perfetto monoatomico." *Rend. Lincei* 3 (1926): 145–49. *FP*, 1:178–85.
31. "Zur Quantelung des idealen einatomigen Gases." *Z. Physik* 36 (1926): 902–12. *FP*, 1:186–95.
32. "Sopra l'intensità delle righe proibite nei campi magnetici intensi." *Rend. Lincei* 3 (1926): 478–83. *FP*, 1:196–200.
33. "Argomenti pro e contro la ipotesi dei quanti di luce." *Nuovo Cimento* 3 (1926): 47–54. *FP*, 1:201–6.
34. "Problemi di chimica, nella fisica dell'atomo." *Periodico di Matematiche* 6 (1926): 19–26. *FP*, 1:207–11.
35. "Sopra l'elettrone rotante." With F. Rasetti. *Nuovo Cimento* 3 (1926): 226–35. *FP*, 1:212–17.
36. "Zur Wellenmechanik des Stossvorganges." *Z. Physik* 40 (1926): 399–402. *FP*, 1:218–21.
37. "Il principio delle adiabatiche e la nozione di forza viva nella nuova meccanica ondulatoria." With E. Persico. *Rend. Lincei* 4 (2) (1926): 452–57. *FP*, 1:222–26.
38a. "Sopra una formula di calcolo delle probabilità." *Nuovo Cimento* 3 (1926): 313–18.
38b. "Un teorema di calcolo delle probabilità ed alcune sue applicazioni." Dissertation for qualification for the Scuola Normale Superiore, Pisa, 1922. *FP*, 1:227–43.
39. "Quantum Mechanics and the Magnetic Moment of Atoms." *Nature* (London) 118 (1926): 876 (letter). *FP*, 1:244–45.
40a. "Eine Messung des Verhältnisses h/k durch die anomale Dispersion des Thalliumdampfes." With F. Rasetti. *Z. Physik* 43 (1927): 379–83.
40b. "Una misura del rapporto h/k per mezzo della dispersione anomala del tallio." With F. Rasetti. *Rend. Lincei* 5 (1927): 566–70. *FP*, 1: 246–50.

Bibliography

41. "Gli effetti elettro e magnetoottici e le loro interpretazioni." *L'Energia Elettrica,* special issue commemorating centenary of death of A. Volta, pp. 109–20. Rome: Uniel, 1927, *FP,* 1:251–70.
42. "Sul meccanismo dell'emissione nella meccanica ondulatoria." *Rend. Lincei* 5 (1927): 795–800. *FP,* 1:271–76.
43. "Un metodo statistico per la determinazione di alcune proprietà dell'atomo." *Rend. Lincei* 6 (1927): 602–7. *FP,* 1:277–82.
44. "Sulla deduzione statistica di alcune proprietà dell'atomo. Applicazione alla teoria del sistema periodico degli elementi." *Rend. Lincei* 7 (1928): 342–46. *FP,* 1:283–86.
45. "Sulla deduzione statistica di alcune proprietà dell'atomo. Calcolo della correzione di Rydberg per i termini s." *Rend. Lincei* 7 (1928): 726–30. *FP,* 1:287–90.
46. "Anomalous Groups in the Periodic System of Elements." *Nature* (London) 121 (1928): 502 (letter).
47. "Eine statistische Methode zur Bestimmung einiger Eigenschaften des Atoms und ihre Anwendung auf die Theorie des periodischen Systems der Elemente." *Z. Physik* 48 (1928): 73–79.
48. "Statistische Berechnung der Rydbergkorrektionen der s-Terme." *Z. Physik* 49 (1928): 550–54.
49. "Über die Anwendung der statistischen Methode auf die Probleme des Atombaues." In *Quantentheorie und Chemie: Leipziger Vorträge,* edited by H. Falkenhagen, pp. 95–111. Leipzig: Hirzel, 1928. *FP,* 1:291–304.
50. "Sopra l'elettrodinamica quantistica." *Rend. Lincei* 9 (1929): 881–87. *FP,* 1:305–10.
51. "Sul moto di un corpo di massa variabile." *Rend. Lincei* 9 (1929): 984–86. *FP,* 1:311–13.
52. "Sulla teoria quantistica delle frange di interferenza." *Rend. Lincei* 10 (1929): 72–77; *Nuovo Cimento* 7 (1930): 153–58. *FP,* 1:314–18.
53. "Sul complesso 4d della molecola di elio." *Rend. Lincei* 10 (1929): 515–17; *Nuovo Cimento* 7 (1930): 159–61. *FP,* 1:319–21.
54a. "Über das Intensitätsverhältnis der Dublettkomponenten der Alkalien." *Z. Physik* 59 (1930): 680–86. *FP,* 1:322–27.
54b. "Sul rapporto delle intensità nei doppietti dei metalli alcalini." *Nuovo Cimento* 7 (1930): 201–7.
55. "Magnetic Moments of Atomic Nuclei." *Nature* (London) 125 (1930): 16 (letter). *FP,* 1:328–29.
56. "I fondamenti sperimentali delle nuove teorie fisiche." *Atti Soc. It. Progr. Sci.,* 18th meeting, 1 (1929): 365–71. *FP,* 1:330–35.
57a. "Sui momenti magnetici dei nuclei atomici." *Mem. Accad. d'Italia* 1 (Physics) (1930): 139–148. *FP,* 1:336–48.
57b. "Über die magnetischen Momente der Atomkerne." *Z. Physik* 60 (1930): 320–33.
58. "Problemi attuali della fisica." *Annali dell'Istruzione media* 5 (1929): 424–28.
59. "L'interpretazione del principio di causalità nella meccanica quan-

Bibliography

tistica." *Rend. Lincei* 11 (1930): 980–85; *Nuovo Cimento* 7 (1930): 361–66. *FP*, 1:349–54.
60. "Atomi e stelle." *Atti Soc. It. Progr. Sci.*, 19th meeting, 1 (1930): 228–35. *FP*, 1: 355–60.
61. "I fondamenti sperimentali della nuova meccanica atomica." *Periodico di Matematiche* 10 (1930): 71–84. *FP*, 1:361–70.
62. "La fisica moderna." *Nuova Antologia* 65 (1930): 137–45. *FP*, 1:371–78.
63. "Sul calcolo degli spettri degli ioni." *Mem. Accad. d'Italia* 1 (Physics) (1930): 149–56; *Nuovo Cimento* 8 (1931): 7-14. *FP*, 1:379–85.
64. "Sopra l'elettrodinamica quantistica." *Rend. Lincei* 12 (1930): 431–35. *FP*, 1:386–90.
65. "Le masse elettromagnetiche nella elettrodinamica quantistica." *Nuovo Cimento* 8 (1931): 121–32. *FP*, 1:391–400.
66. "La théorie du rayonnement." *Annales de l'Inst. H. Poincaré* 1 (1931): 53–74.
67. "Quantum Theory of Radiation." *Rev. Mod. Phys.* 4 (1932): 87–132. *FP*, 1:401–45.
68. "Über den Ramaneffekt des Kohlendioxyds." *Z. Physik* 71 (1931): 250–59. *FP*, 1:446–54.
69. "Über den Ramaneffekt des Steinsalzes." With F. Rasetti. *Z. Physik* 71 (1931): 689–95. *FP*, 1:455–60.
70. "Über die Wechselwirkung von zwei Elektronen." With H. Bethe. *Z. Physik* 77 (1932): 296–306. *FP*, 1:461–71.
71. "L'effetto Raman nelle molecole e nei cristalli." *Mem.Accad. d'Italia* 3 (Physics) (1932): 239–56. *FP*, 1:472–87.
72a. "État actuel de la physique du noyau atomique." Fifth International Congress on Electricity, Paris, 1932, Comptes Rendus. 1 sect., rep. 22, 789–807.
72b. "Lo stato attuale della fisica del nucleo atomico." *Ric. Scientifica* 3 (2) (1932): 101–13. *FP*, 1:488–502.
73. "Sulle bande di oscillazione e rotazione dell'ammoniaca." *Rend. Lincei* 16 (1932): 179–85; *Nuovo Cimento* 9 (1932): 277–83. *FP*, 1:503–8.
74. "Azione del campo magnetico terrestre sulla radiazione penetrante." With B. Rossi. *Rend. Lincei* 17 (1933): 346–50; see also *Nuovo Cimento* 10 (1933): 333–38. *FP*, 1:509–13.
75a. "Zur Theorie der Hyperfeinstrukturen." With E. Segrè. *Z. Physik* 82 (1933): 11–12, 729–49.
75b. "Sulla teoria delle strutture iperfini." With E. Segrè. *Mem. Accad. d'Italia* 4 (Physics) (1933): 131–58. *FP*, 1:514–37.
76. "Tentativo di una teoria dell'emissione dei raggi *beta*." *Ric. Scientifica* 4 (2) (1933): 491–95. *FP*, 1:538–44.
77a. "Sulla ricombinazione di elettroni e positroni." With G. Uhlenbeck. *Ric. Scientifica* 4 (2) (1933): 157–60.
77b. "On the Recombination of Electrons and Positrons." With G. Uhlenbeck. *Phys. Rev.* 44 (1933): 510–11. *FP*, 1:545–47.

Bibliography

78. "Uno spettrografo per raggi *gamma* a cristallo di bismuto." With F. Rasetti. *Ric. Scientifica* 4 (2) (1933): 299–302. *FP*, 1:548–52.
79. "Le ultime particelle costitutive della materia." *Atti Soc. It. Progr. Sci.*, 22d meeting, 2 (1933): 7–14; *Scientia* 55 (1934): 21–28. *FP*, 1:553–58.
80a. "Tentativo di una teoria dei raggi β." *Nuovo Cimento* 11 (1934): 1–19. *FP*, 1:559–74.
80b. "Versuch einer Theorie der β-Strahlen. 1." *Z. Physik* 88 (1934): 161–71. *FP*, 1:575–90.
81. "Zur Bemerkung von G. Beck und K. Sitte." *Z. Physik* 89 (1934): 522. *FP*, 1:591.
82. "Le orbite ∞-s degli elementi." With E. Amaldi. *Mem. Accad. d'Italia* 6, no. 1 (Physics) (1934): 119–49. *FP*, 1:592–619.
83. "Statistica, meccanica." In *Enciclopedia Italiana di Scienze, Lettere, ed Arti*, 32: 518–23. Rome: Istituto G. Treccani, 1936. *FP*, 1:620–38.
84a. "Radioattivitià provocata da bombardamento di neutroni I." *Ric. Scientifica* 5 (1) (1934): 283. *FP*, 1:639–46.
84b. "Radioactivity Induced by Neutron Bombardment I." Translation issued by John Crerar Library. *Ric. Scientifica* 5 (1) (1934): 283. *FP*, 1:674–75.
85a. "Radioattività provocata da bombardamento di neutroni II." *Ric. Scientifica* 5 (1) (1934): 330–31. *FP*, 1:647–48.
85b. "Radioactivity Produced by Neutron Bombardment II." Translation issued by John Crerar Library. *Ric. Scientifica* 5 (1) (1934): 330–31. *FP*, 1:676.
86a. "Radioattività provocata da bombardamento di neutroni III." With E. Amaldi, O. D'Agostino, F. Rasetti, and E. Segrè. *Ric. Scientifica* 5 (1) (1934): 452–53. *FP*, 1:649–50.
86b. "Beta-Radioactivity Produced by Neutron Bombardment III." Translation issued by John Crerar Library. With E. Amaldi, O. D'Agostino, F. Rasetti, and E. Segrè. *Ric. Scientifica* 5 (1) (1934): 452–53. *FP*, 1:677–8.
87a. "Radioattività provocata da bombardamento di neutroni IV." With E. Amaldi, O. D'Agostino, F. Rasetti, and E. Segrè. *Ric. Scientifica* 5 (1) (1934): 652–53. *FP*, 1:651–52.
87b. "Radioactivity Produced by Neutron Bombardment. IV." Translation issued by John Crerar Library. With E. Amaldi, O. D'Agnostino, F. Rasetti, and E. Segrè. *Ric. Scientifica* 5 (1) (1934): 652–53. *FP*, 1:679–80.
88a. "Radioattività provocata da bombardamento di neutroni V." With E. Amaldi, O. D'Agostino, F. Rasetti, and E. Segrè. *Ric. Scientifica* 5 (2) (1934): 21–22. *FP*, 1:653–54.
88b. "Radioactivity Produced by Neutron Bombardment V." Translation issued by John Crerar Library. With E. Amaldi, O. D'Agostino, F. Rasetti, and E. Segrè. *Ric. Scientifica* 5 (2) (1934): 21–22. *FP*, 681–2.
89a. "Radioattività provocata da bombardamento di neutroni VII."

Bibliography

With E. Amaldi, O. D'Agostino, B. Pontecorvo, F. Rasetti, and E. Segrè. *Ric. Scientifica* 5 (2) (1934): 467–70. *FP*, 1:655–60.

89b. "Radioactivity Produced by Neutron Bombardment VII." Translation issued by John Crerar Library. With E. Amaldi, O. D'Agostino, B. Pontecorvo, F. Rasetti, and E. Segrè. *Ric. Scientifica* 5 (2) (1934): 467–70. *FP*, 1:683–88.

90a. "Radioattività provocata da bombardamento di neutroni VIII." With E. Amaldi, O. D'Agostino, B. Pontecorvo, F. Rasetti, and E. Segrè. *Ric. Scientifica* 6, no. 1 (1935): 123–25. *FP*, 1:661–64.

90b. "Radioactivity Produced by Neutron Bombardment VIII." Translation issued by John Crerar Library. With E. Amaldi, O. D'Agostino, B. Pontecorvo, F. Rasetti, and E. Segrè. *Ric. Scientifica* 6 (1) (1935): 123–25. *FP*, 1:689–92.

91a. "Radioattività provocata da bombardamento di neutroni IX." With E. Amaldi, O. D'Agostino, B. Pontecorvo, and E. Segrè. *Ric. Scientifica* 6 (1) (1935): 435–37. *FP*, 1:665–68.

91b. "Radioactivity Produced by Neutron Bombardment IX." Translation issued by John Crerar Library. With E. Amaldi, O. D'Agostino, B. Pontecorvo, and E. Segrè. *Ric. Scientifica* 6 (1) (1935): 435–37. *FP*, 1:693–96.

92a. "Radioattività provocata da bombardamento di neutroni X." With E. Amaldi, O. D'Agostino, B. Pontecorvo, and E. Segrè. *Ric. Scientifica* 6, no. 1 (1935): 581–84. *FP*, 1:669–73.

92b. "Radioactivity Produced by Neutron Bombardment X." Translation issued by John Crerar Library. With E. Amaldi, O. D'Agostino, B. Pontecorvo, and E. Segrè. *Ric. Scientifica* 6 (1) (1935): 581–84. *FP*, 1:697–701.

93. "Radioactivity Induced by Neutron Bombardment." *Nature* (London) 133 (1934): 757 (letter). *FP*, 1:702–3.

94. "Sulla possibilità di produrre elementi di numero atomico maggiore di 92." With F. Rasetti and O. D'Agostino. *Ric. Scientifica* 5 (1) (1934): 536–37. *FP*, 1:704–5.

95. "Sopra lo spostamento per pressione delle righe elevate delle serie spettrali." *Nuovo Cimento* 2 (1934): 157–66. *FP*, 1:706–14.

96. "Radioattività prodotta da bombardamento di neutroni." *Nuovo Cimento* 2 (1934): 429–41. *FP*, 1:715–24.

97. "Nuovi radioelementi prodotti con bombardamento di neutroni." With E. Amaldi, F. Rasetti, and E. Segrè. *Nuovo Cimento* 2 (1934): 442–51. *FP*, 1:725–31.

98. "Artificial Radioactivity Produced by Neutron Bombardment." With E. Amaldi, O. D'Agostino, F. Rasetti and E. Segrè. *Proc. Roy. Soc.* (London), Series A, 146 (1934): 483–500. *FP*, 1:732–47.

99. "Possible Production of Elements of Atomic Number Higher than 92." *Nature* (London) 133 (1934): 898–99. *FP*, 1:748–50.

100. "Artificial Radioactivity Produced by Neutron Bombardment." *Nature* (London) 134 (1934): 668. *FP*, 1:751.

101. "Conferencias." *Facultad de ciencias exactas fisicas y naturales*, Publicación 15. Buenos Aires, 1934.

Bibliography

102. "Natural Beta Decay." In *International Conference on Physics*, vol. 1, *Nuclear Physics*, pp. 66–71. London: Physical Society. 1934. *FP*, 1:752–3.
103. "Artificial Radioactivity Produced by Neutron Bombardment." In *International Conference on Physics*, vol. 1, *Nuclear Physics*, pp. 75–77. London, 1934. *FP*, 1:754–56.
104. "La radioattività artificiale." *Atti. Soc. It Progr. Sci.*, 23d meeting, 1 (1934): 34–39.
105a. "Azione di sostanze idrogenate sulla radioattività provocata da neutroni. I." With E. Amaldi, B. Pontecorvo, F. Rasetti, and E. Segrè. *Ric. Scientifica* 5 (2) (1934): 282–83. *FP*, 1:757–58.
105b. "Influence of Hydrogenous Substances on the Radioactivity Produced by Neutrons. I" With E. Amaldi, B. Pontecorvo, F. Rasetti, and E. Segrè. Translated by E. Segrè, *FP*, 1:761–62.
106a. "Effetto di sostanze idrogenate sulla radioattività provocata da neutroni. II." With B. Pontecorvo and F. Rasetti. *Ric. Scientifica* 5 (2) (1934): 380–81. *FP*, 1:759–60.
106b. "Influence of Hydrogenous Substances on the Radioactivity Produced by Neutrons. II" With B. Pontecorvo and F. Rasetti. Translated by F. Rasetti, *FP*, 1:763–64.
107. "Artificial Radioactivity Produced by Neutron Bombardment. Part II." With E. Amaldi, O. D'Agostino, B. Pontecorvo, F. Rasetti and E. Segrè. *Proc. Roy. Soc.* (London), Series A, 149 (1935): 522–58. *FP*, 1:765–94.
108. "Ricerche sui neutroni lenti." With F. Rasetti. *Nuovo Cimento* 12 (1935): 201–10. *FP*, 1:795–802.
109. "On the Velocity Distribution Law for the Slow Neutrons." In *Verhandelingen, op 25 Mei 1935 aangeboden aan Prof. Dr. P. Zeeman*, pp. 128–30. The Hague: Martinus Nijhoff, 1935. *FP*, 1:803–5.
110. "On the Recombination of Neutrons and Protons." *Phys. Rev.* 48 (1935): 570. *FP*, 1:806–7.
111. "Recenti risultati della radioattività artificiale." *Ric. Scientifica* 6 (2) (1935): 399–402; *Atti Soc. It. Progr. Sci.*, 24th meeting, 3 (1935): 116–20.
112. "Sull'assorbimento dei neutroni lenti. I." With E. Amaldi. *Ric. Scientifica* 6 (2) (1935): 344–47. *FP*, 1:808–15.
113. "——— II." With E. Amaldi. *Ric. Scientifica* 6 (2) (1935): 443–47. *FP*, 1:816–22.
114. "——— III." With E. Amaldi. *Ric. Scientifica* 7 (1) (1936): 56–59. *FP*, 1:823–27.
115. "Sul cammino libero medio dei neutroni nella paraffina." With E. Amaldi. *Ric. Scientifica* 7 (1) (1936): 223–25. *FP*, 1:828–31.
116. "Sui gruppi di neutroni lenti." With E. Amaldi. *Ric. Scientifica* 7 (1) (1936): 310–15. *FP*, 1:832–36.
117. "Sulle proprietà di diffusione dei neutroni lenti." With E. Amaldi. *Ric. Scientifica* 7 (1) (1936): 393–95. *FP*, 1:837–40.
118a. "Sopra l'assorbimento e la diffusione dei neutroni lenti." With E.

Bibliography

- Amaldi. *Ric. Scientifica* 7 (1) (1936): 454–503. *FP*, 1:841–91.
- 118b. "On the Absorption and the Diffusion of Slow Neutrons." With E. Amaldi. *Phys. Rev.* 50 (1936): 899–928. *FP*, 1:892–942.
- 119a. "Sul moto dei neutroni nelle sostanze idrogenate." *Ric. Scientifica* 7 (2) (1936): 13–52. *FP*, 1:943–79.
- 119b. "On the Motion of Neutrons in Hydrogenous Substances." Translation of 119a by G. Temmer. *FP*, 1:980–1016.
- 120. "Un maestro: Orso Mario Corbino." *Nuova Antologia* 72 (1937): 313–16; see also *L'Energia Elettrica* 14 (1937): 85–6. *FP*, 1:1017–20.
- 121. "Un generatore artificiale di neutroni." With E. Amaldi and F. Rasetti. *Ric. Scientifica* 8 (2) (1937): 40–43. *FP*, 1:1021–24.
- 122. "Neutroni lenti e livelli energetici nucleari." *Nuovo Cimento* 15 (1938): 41–42. (Abridged). *FP*, 1:1025–26.
- 123. "Tribute to Lord Rutherford." *Nature* (London) 140 (1937): 1052. *FP*, 1:1027.
- 124. "Azione del boro sui neutroni caratteristici dello iodio." With E. Rasetti. *Ric. Scientifica* 9 (2) (1938): 472–73. *FP*, 1:1028–29.
- 125. "On the Albedo of Slow Neutrons." With E. Amaldi and G. C. Wick. *Phys. Rev.* 53 (1938): 493. *FP*, 1:1030–31.
- 126. "Prospettive di applicazioni della radioattività artificiale." *Rendiconti dell'Istituto di Sanità Pubblica* 1 (1938): 421–32.
- 127. "Guglielmo Marconi e la propagazione delle onde elettromagnetiche nell'alta atmosfera." *Soc. It. Progr. Sci., Collectanea Marconiana*, pp. 1–5. Rome, 1938. *FP*, 1:1032–36.
- 128. "Artificial Radioactivity Produced by Neutron Bombardment." *Les Prix Nobel en 1938*, pp. 1–8. Stockholm, 1939. *FP*, 1:1037–43.
- 129. "The Fission of Uranium." With H. L. Anderson, E. T. Booth, J. R. Dunning, G. N. Glasoe, and F. G. Slack. *Phys. Rev.* 55 (1939): 511–12 (letter). *FP*, 2:1–4.
- 130. "Production of Neutrons in Uranium Bombarded by Neutrons." With H. L. Anderson and H. B. Hanstein. *Phys. Rev.* 55 (1939): 797–98 (letter). *FP*, 2:5–7.
- 131. "Simple Capture of Neutrons by Uranium." With H. L. Anderson. *Phys. Rev.* 55 (1939): 1106–7 (letter) *FP*, 2:8–10.
- 132. "Neutron Production and Absorption in Uranium." With H. L. Anderson and L. Szilard. *Phys. Rev.* 56 (1939): 284–86. *FP*, 2:11–14.
- 133. "The Absorption of Mesotrons in Air and in Condensed Materials." *Phys. Rev.* 56 (1939): 1242 (letter). *FP*, 1:15–17.
- 134. "The Ionization Loss of Energy in Gases and in Condensed Materials." *Phys. Rev.* 57 (1940): 485–93. *FP*, 2:18–28.
- 135. "Fission of Uranium by Alpha-Particles." With E. Segrè. *Phys. Rev.* 59 (1941): 680–81 (letter). *FP*, 2:29–30.
- 136. "Production and Absorption of Slow Neutrons by Carbon." With H. L. Anderson. Metallurgical Laboratory Report A–21, September 25, 1940. *FP*, 2:31–40.
- 137. "Branching Ratios in the Fission of Uranium (235)." With H. L.

Bibliography

Anderson and A. V. Grosse. *Phys. Rev.* 59 (1941): 52–56. *FP*, 2:41–49.
137a. "Reactions Produced by Neutrons in Heavy Elements." *Nature* 146 (1940): 640–42; see also *Science* 92 (1940): 269–71.
138. "Production of Neutrons by Uranium." With H. L. Anderson. Metallurgical Laboratory Report A–6, January 17, 1941. *FP*, 2:50–69.
139. "Capture of Resonance Neutrons by a Uranium Sphere Imbedded in Graphite." With H. L. Anderson, R. R. Wilson, and E. C. Creutz. Appendix A of Metallurgical Laboratory Report A–12 to The National Defense Research Committee by H. D. Smyth, Princeton University, June 1, 1941. *FP*, 2:70–75.
140. "Standards in Slow Neutron Measurements." With H. L. Anderson. Metallurgical Laboratory Report A–2, June 5, 1941. *FP*, 2:76–85.
141. "Some Remarks on the Production of Energy by a Chain Reaction in Uranium." Metallurgical Laboratory Report A–14, June 30, 1941. *FP*, 2:86–90.
142. "The Absorption of Thermal Neutrons by a Uranium Sphere Imbedded in Graphite." With G. L. Weil. Metallurgical Laboratory Report A–1, July 3, 1941. *FP*, 2:91–97.
143. "Remarks on Fast Neutron Reactions." Metallurgical Laboratory Report A–46, October 6, 1941. *FP*, 2:98–103.
144. "The Effect of Chemical Binding in the Scattering and Moderation of Neutrons by Graphite." Metallurgical Laboratory Report C–87, October (?) 1941. *FP*, 2:104–6.
145. "Fission Cross Section of Unseparated Uranium for Fast Rn + Be Neutrons." With H. L. Anderson. Metallurgical Laboratory Report C–83, November (?) 1941. *FP*, 2:108–8.
146. "Absorption Cross Sections for Rn + Be Fast Neutrons." With H. L. Anderson and G. L. Weil. Metallurgical Laboratory Report C–72, November (?) 1941. *FP*, 2:109–11.
147. "Neutrons Emitted by a Ra + Be Photosource." With B. T. Feld. Metallurgical Laboratory Report CP–89. *FP*, 2:112–15.
148. "The Absorption Cross Section of Boron for Thermal Neutrons." With H. L. Anderson. Metallurgical Laboratory Report CP–74, January 1942. *FP*, 2:116–18.
149. "Neutron Production in a Lattice of Uranium and Graphite." (Theoretical Part). Metallurgical Laboratory Report CP–12, March 17, 1942. *FP*, 2:119–27.
150. "Neutron Production in a Lattice of Uranium Oxide and Graphite." (Exponential Experiment). With H. L. Anderson, B. T. Feld, G. L. Weil, and W. H. Zinn. Metallurgical Laboratory Report CP–20, March 26, 1942. *FP*, 2:128–36.
151. "Preliminary Report on the Exponential Experiment at Columbia University." Metallurgical Laboratory Report CP–26, March–April, 1942. *FP*, 2:137–43.
152. "Effect of Atmospheric Nitrogen and of Changes of Temperature

Bibliography

on the Reproduction Factor." Metallurgical Laboratory Report CP–85, May 19, 1942. *FP*, 2:144–46.

153. "A Table for Calculating the Percentage of Loss Due to the Presence of Impurities in Alloy." Metallurgical Laboratory Report C–5, February 10, 1942. *FP*, 2:147–8.
154. "The Temperature Effect on a Chain Reacting Unit. Effect of the Change of Leakage." Metallurgical Laboratory Report C–8, February 25, 1942. *FP*, 2:149–51.
155. "The Use of Reflectors and Seeds in a Power Plant." With G. Breit. Metallurgical Laboratory Report C-11, March 9, 1942. *FP*, 2:152–58.
156. "Slowing Down and Diffusion of Neutrons." Metallurgical Laboratory Report C–29, notes on lecture of March 10, 1942. *FP*, 2:159–64.
157. "Determination of the Albedo and the Measurement of Slow Neutron Density." Metallurgical Laboratory Report C–31, notes on lecture of March 17, 1942. *FP*, 2:165–69.
158. "The Number of Neutrons Emitted by a Ra + Be Source (Source I)." With H. L. Anderson, J. H. Roberts, and M. D. Whitaker, Metallurgical Laboratory Report C–21, March 21, 1942. *FP*, 2:170–74.
159. "The Determination of the Ratio Between the Absorption Cross Sections of Uranium and Carbon for Thermal Neutrons." Metallurgical Laboratory Report C–84, May 15, 1942. *FP*, 175–76.
160. "The Absorption of Graphite for Thermal Neutrons." Metallurgical Laboratory Report C–154, notes on lecture of June 30, 1942. *FP*, 2:177–78.
161. "Longitudinal Diffusion in Cylindrical Channels." With A. M. Weinberg. Metallurgical Laboratory Report C–170, July 7, 1942. *FP*, 2:179–85.
162. "The Number of Neutrons Emitted by Uranium per Thermal Neutron Absorbed." Metallurgical Laboratory Report C–190, July 16, 1942. *FP*, 2:186–94.
163. "Effect of Temperature Changes on the Reproduction Factor." With R. F. Christy and A. M. Weinberg. Metallurgical Laboratory Report CP–254, September 14, 1942. *FP*, 2:195–99.
164. "Status of Research Problems in Experimental Nuclear Physics." Excerpt from Metallurgical Laboratory Report C–133 for week ending June 20, 1942. *FP*, 2:200–202.
165. "Status of Research Problems in Experimental Nuclear Physics." Excerpt from Metallurgical Laboratory Report C–207 for week ending July 25, 1942. *FP*, 2:203–5.
166. "Status of Research Problems of the Physics Division." Excerpt from Metallurgical Laboratory Report CP–235 for month ending August 15, 1942. *FP*, 2:206–9.
167. "Exponential Pile No. 2." Excerpt from Metallurgical Laboratory Report CA–247 for week ending August 29, 1942. *FP*, 2:210–11.

Bibliography

168. "Status of Research Problems of the Physics Division." Excerpt from Metallurgical Laboratory Report CP–257 for month ending September, 15, 1942. *FP*, 2:212–15.
169. "Purpose of the Experiment at the Argonne Forest. Meaning of the Reproduction Factor "k"." Metallurgical Laboratory Report CP–283, notes on lecture of September 23, 1942. *FP*, 2:216–18.
170. "The Critical Size—Measurement of "k" in the Exponential Pile." Metallurgical Laboratory Report CP–289, notes on lecture of September 30, 1942. *FP*, 2:219–25.
171. "Problem of Time Dependence of the Reaction Rate: Effect of Delayed Neutrons Emission." Metallurgical Laboratory Report CP–291, notes on lecture of October 7, 1942. *FP*, 2:226–29.
172. "A Simplified Control. Optimum Distribution of Materials in the Pile." Metallurgical Laboratory Report CP–314, notes on lecture of October 20, 1942. *FP*, 2:230–33.
173. "Design of the Graphite-Uranium Lattice: Experimental Determination of f_t from the Cd Ratio." Metallurgical Laboratory Report CP–337, notes on lectures of October 27 and November 3, 1942. *FP*, 2:234–39.
174. "Calculation of the Reproduction Factor." Metallurgical Laboratory Report CP–358, notes on lecture of November 10, 1942. *FP*, 2:240–42.
175. "The Projected Experiment at Argonne Forest and the Reproduction Factor in Metal Piles." Excerpt from Metallurgical Laboratory Report CP–297 for month ending October 15, 1942. *FP*, 2:243–47.
176. "Methods of Cooling Chain Reacting Piles." Metallurgical Laboratory (CP) Memo–10, October 5, 1942, *FP*, 2:248–54.
177. "The Effect of Bismuth on the Reproduction Factor." Excerpt from Metallurgical Laboratory Report CA–320, Bulletin for week ending October 31, 1942. *FP*, 2:255–56.
178. "The Experimental Chain Reacting Pile and Reproduction Factor in Some Exponential Piles." Excerpt from Metallurgical Laboratory Report CP–341 for month ending November 15, 1942. *FP*, 2:257–62.
179. "Feasibility of a Chain Reaction." Metallurgical Laboratory Report CP–383, November 26, 1942. *FP*, 2:263–67.
180. "Work Carried out by the Physics Division." Excerpt from Metallurgical Laboratory Report CP–387 for month ending December 15, 1942. *FP*, 2:268–71.
181. "Experimental Production of a Divergent Chain Reaction." *Am. J. Phys.* 20 (1952): 536–58. *FP*, 2:272–307.
182. "Summary of Experimental Research Activities." Excerpt from Metallurgical Laboratory Report CP–416 for month ending January 15, 1943. *FP*, 2:308–10.
183. "Summary of Experimental Research Activities." Excerpt from Metallurgical Laboratory Report CP–455 for month ending February 6, 1943. *FP*, 2:311–14.

Bibliography

184. "The Utilization of Heavy Hydrogen in Nuclear Chain Reactions." Memorandum of conference with Prof. H. C. Urey on March 6, 7, and 8, 1943. Metallurgical Laboratory Report A–544. *FP*, 2:315–19.
185. "The Slowing Down of Neutrons in Heavy Water." Metallurgical Laboratory Report CP–530, March 19, 1943. *FP*, 2:320–23.
186. "Summary of Experimental Research Activities." Excerpt from Metallurgical Laboratory Report CP–570 for month ending April 17, 1943. *FP*, 2:324–25.
187. "Summary of Experimental Research Activities." Excerpt from Metallurgical Laboratory Report CP–641 for month ending May 10, 1943. *FP*, 2:326–27.
188. "Standardization of the Argonne Pile." With H. L. Anderson, J. Marshall, and L. Woods. Excerpt from Metallurgical Laboratory Report CP–641 for month ending May 10, 1943. *FP*, 2:328–32.
189. "Tests on a Shield for the Pile at Site W." With W. H. Zinn. Metallurgical Laboratory Report CP–684, May 25, 1943. *FP*, 2:333–34.
190. "Summary of Experimental Research Activities." Excerpt from Metallurgical Laboratory Report CP–718 for month ending June 12, 1943, *FP*, 2:335–36.
191. "Production of Low Energy Neutrons by Filtering through Graphite." With H. L. Anderson and L. Marshall. *Phys. Rev.* 70 (1946): 815–17. *FP*, 2:337–41.
192. "Summary of Experimental Research Activities." Excerpt from Metallurgical Laboratory Report CP–781 for month ending July 10, 1943. *FP*, 2:342–44.
193. "Range of Indium Resonance Neutrons from a Source of Fission Neutrons." With G. L. Weil. Excerpt from Metallurgical Laboratory Report CP–871 for month ending August 14, 1943. *FP*, 2:345–46.
194. "Summary of Experimental Research Activities." Excerpt from Metallurgical Laboratory Report CP–1016 for month ending October 23, 1943. *FP*, 2:347–48.
195. "Summary of Experimental Research Activities." Excerpt from Metallurgical Laboratory Report CP–1088 for month ending November 23, 1943. *FP*, 2:349–50.
196. "The Range of Delayed Neutrons." With G. Thomas. Excerpt from Metallurgical Laboratory Report CP–1088 for month ending November 23, 1943. *FP*, 2:351–52.
197. "Slowing Down of Fission Neutrons in Graphite." With J. Marshall and L. Marshall. Metallurgical Laboratory Report CP–1084, November 25, 1943. *FP*, 2:353–57.
198. "Fission Cross-Section and ν-Value for 25." With J. Marshall and L. Marshall. Metallurgical Laboratory Report CP–1186, December 31, 1943. *FP*, 2:358–63.
199. "Summary of Experimental Research Activities." Excerpt from

Bibliography

Metallurgical Laboratory Report CP-1175 for month ending December 25, 1943. *FP*, 2:364-65.
200. "A Thermal Neutron Velocity Selector and its Application to the Measurement of the Cross-Section of Boron." With J. Marshall and L. Marshall. *Phys. Rev.* 72 (1947): 193-96. *FP*, 2:366-71.
201. "Summary of Experimental Research Activities." Excerpt from Metallurgical Laboratory Report CP-1255 for month ending January 24, 1944. *FP*, 2:372-75.
202. "Summary of Experimental Research Activities." Excerpt from Metallurgical Laboratory Report CP-1389 for month ending February 24, 1944. *FP*, 2:376-79.
203. "Report of Fermi's Activities with the Marshall Group." Excerpt from Metallurgical Laboratory Report CP-1389 for month ending February 24, 1944. *FP*, 2:380-82.
204. "Summary of Experimental Research Activities." Excerpt from Metallurgical Laboratory Report CP-1531 for month ending March 25, 1944. *FP*, 2:383-84.
205. "Range of Fission Neutrons in Water." With H. L. Anderson and D. E. Nagle. Excerpt from Metallurgical Laboratory Report CP-1531 for month ending March 25, 1944. *FP*, 2:385-87.
206. "Evidence for the Formation of 26." With L. Marshall. Excerpt from Metallurgical Laboratory Report CP-1531 for month ending March 25, 1944. *FP*, 2:388-90.
207. "Summary of Experimental Research Activities." Excerpt from Metallurgical Laboratory Report CP-1592 for month ending April 24, 1944. *FP*, 2:391-93.
208. "Absorption of 49." Excerpt from Metallurgical Laboratory Report CP-1592 for month ending April 24, 1944. *FP*, 2:394.
209. "Comparison of the Ranges in Graphite of Fission Neutrons from 49 and 25." With A. Heskett and D. E. Nagle. Excerpt from Metallurgical Laboratory Report CP-1592 for month ending April 24, 1944. *FP*, 2:395-96.
210. "Method for Measuring Neutron-Absorption Cross Sections by the Effect on the Reactivity of a Chain-Reacting Pile." With H. L. Anderson, A. Wattenberg, G. L. Weil, and W. H. Zinn. *Phys. Rev.* 72 (1947): 16-23. *FP*, 2:397-410.
211. "Discussion on Breeding." Excerpt from Metallurgical Laboratory Report N-1729, notes on meeting of April 26, 1944. *FP*, 2:411-14.
212. "Report on Recent Values of Constants of 25 and 49." Metallurgical Laboratory Report CK-1788, May 19, 1944. *FP*, 2:415-18.
213. "Summary of Experimental Research Activities." Excerpt from Metallurgical Laboratory Report CP-1729 for month ending May 25, 1944. *FP*, 2:419-20.
214. "Dissociation Pressure of Water Due to Fission." With H. L. Anderson. Excerpt from Metallurgical Laboratory Report CP-1729 for month ending May 25, 1944. *FP*, 2:421-22.
215. "Summary of Experimental Research Activities." Excerpt from

Bibliography

Metallurgical Laboratory Report CK–1761 for month ending May 25, 1944. *FP*, 2:423.

216. "Summary of Experimental Research Activities." Excerpt from Metallurgical Laboratory Report CP–1827 for month ending June 25, 1944. *FP*, 2:424.
217. "Collimation of Neutron Beam from Thermal Column of CP–3 and the Index of Refraction for Thermal Neutrons." With W. H. Zinn. Excerpt from Metallurgical Laboratory Report CP–1965 for month ending July 29, 1944. *FP*, 2:425–27.
218. "Methods for Analysis of Helium Circulating in the 105 Unit." Document HW 3–492, August 7, 1944. *FP*, 2:428–29.
219. "Boron Absorption of Fission Activation." With H. L. Anderson. Excerpt from Metallurgical Laboratory Report CF–2161 for month ending September 23, 1944. *FP*, 2:430–32.
220. "Reflection of Neutrons on Mirrors." With W. H. Zinn. Physical Society Cambridge (England) Conference Report 1947, p. 92. *FP*, 2:433–34.
221. "Relation of Breeding to Nuclear Properties." Excerpt from Metallurgical Laboratory Report CF–3199, discussion on breeding, June 19–20, 1945. *FP*, 2:435–39.
222. "A Course in Neutron Physics." Part 1. Los Alamos Document LADC–255, February 5, 1946. Notes by I. Halpern. Part 2, declassified in 1962. *FP*, 2: 440–541.
223. "The Development of the First Chain Reacting Pile." *Proc. Amer. Philosophical Soc.* 90 (1946): 20–24. *FP*, 2:542–49.
224. "Atomic Energy for Power." George Westinghouse Centennial Forum, vol. 1. *Science and Civilization—The Future of Atomic Energy*. New York: McGraw-Hill, 1946. Also MDDC–1, Atomic Energy Commission. *FP*, 2:550–57.
225. "Elementary Theory of the Chain-Reacting Pile." *Science* 105 (1947): 27–32. *FP*, 2:558–67.
226. "The Transmission of Slow Neutrons through Microcrystalline Materials." With W. J. Sturm and R. G. Sachs. *Phys. Rev.* 71 (1947): 589–94. *FP*, 2:568–77.
227. "Phase of Neutron Scattering." With L. Marshall. Physical Society Cambridge (England) Conference Report, 1947, pp. 94–97. *FP*, 2:578–82.
228. "Interference Phenomena of Slow Neutrons." With L. Marshall. *Phys. Rev.* 71 (1947): 666–77. *FP*, 2:583–601.
229. "Phase of Scattering of Thermal Neutrons by Aluminum and Strontium." With L. Marshall. *Phys. Rev.* 71 (1947): 915 (letter). *FP*, 2:602–3.
230. "Spin Dependence of Scattering of Slow Neutrons by Be, Al and Bi." With L. Marshall. *Phys. Rev.* 72 (1947): 408–10. *FP*, 2:604–7.
  1. "Further Experiments with Slow Neutrons." With L. Marshall. Excerpts from quarterly reports CF–3574, Argonne National

Bibliography

Laboratory, July 26, 1946, and CP–3750 and CP–3801, Argonne National Laboratory, January 17 and April 14, 1947. *FP,* 2:608–14.

232. "The Decay of Negative Mesotrons in Matter." With E. Teller and V. Weisskopf. *Phys. Rev.* 71 (1947): 314–15. *FP,* 2:615–17.

233. "The Capture of Negative Mesotrons in Matter." With E. Teller. *Phys. Rev.* 72 (1947): 399–408. *FP,* 2:618–32.

234. "On the Interaction Between Neutrons and Electrons." With L. Marshall. *Phys. Rev.* 72 (1947): 1139–46. *FP,* 2:633–45.

235. "Spin Dependence of Slow Neutron Scattering by Deuterons." With L. Marshall. *Phys. Rev.* 75 (1949): 578–80. *FP,* 2:646–51.

236. "Note on Census-Taking in Monte-Carlo Calculations." With R. D. Richtmyer. Los Alamos Document LAMS–805, July 11, 1948. *FP,* 2:652–54.

237. "On the Origin of the Cosmic Radiation." *Phys. Rev.* 75 (1949): 1169–74. *FP,* 2:655–66.

238. "An Hypothesis on the Origin of the Cosmic Radiation." *Nuovo Cimento* 6, suppl. (1949): 317–23. *FP,* 2:667–72.

239. "Are Mesons Elementary Particles?" With C. N. Yang. *Phys. Rev.* 76 (1949): 1739–43. *FP,* 2:673–83.

240. "Conferenze di Fisica Atomica." (Fondazione Donegani). Accademia Nazionale dei Lincei, 1950. *FP,* 2:684–788.

240a. "La visita di Enrico Fermi al Consiglio Nazionale delle Ricerche." *Ric. Scientifica* 19 (1949): 1113–18.

241. "High Energy Nuclear Events." *Prog. Theor. Phys.* 5 (1950): 570–83. *FP,* 2:789–803.

242. "Angular Distribution of the Pions Produced in High Energy Nuclear Collisions." *Phys. Rev.* 81 (1951): 683–87. *FP,* 2:804–12.

243. "Excerpt from a Lecture on Taylor Instability. Given during the Fall of 1951 at Los Alamos Scientific Laboratory." *FP,* 2:813–15.

244. "Taylor Instability of an Incompressible Liquid." Part 1 of Los Alamos Document AECU–2979, 1951. *FP,* 2:816–20.

245. "Taylor Instability at the Boundary of Two Incompressible Liquids." With J. von Neumann. Part 2 of Los Alamos Document AECU–2979. *FP,* 2:821–4.

246. "Fundamental Particles." Proceedings of the International Conference on Nuclear Physics and the Physics of Fundamental Particles. The University of Chicago, September 17–22, 1951 (lecture). *FP,* 2:825–8.

247. "The Nucleus." *Physics Today* 5 (March 1952): 6–9. *FP,* 2:829–34.

248. "Total Cross Sections of Negative Pions in Hydrogen." With H. L. Anderson, E. A. Long, R. Martin, and D. E. Nagle. *Phys. Rev.* 85 (1952): 934–35 (letter). *FP,* 2:835–37.

249. "Ordinary and Exchange Scattering of Negative Pions by Hydrogen." With H. L. Anderson, A. Lundby, D. E. Nagle, and G. B. Yodh. *Phys. Rev.* 85 (1952): 935–36 (letter). *FP,* 2:838–40.

Bibliography

249a. "Scattering of Negative Pions by Hydrogen." With A. Lundby, H. L. Anderson, D. E. Nagle and G. Yodh. (Abstract). *Phys. Rev.* 86 (1952): 603.
250. "Total Cross Sections of Positive Pions in Hydrogen." With H. L. Anderson, E. A. Long, and D. E. Nagle. *Phys. Rev.* 85 (1952): 936 (letter). *FP*, 2:841–43.
251. "Letter to Feynman" (1952). *FP*, 2:844–46.
252. "Deuterium Total Cross Sections for Positive and Negative Pions." With H. L. Anderson, D. E. Nagle, and G. B. Yodh. *Phys. Rev.* 86 (1952): 413 (letter). *FP*, 2:847–49.
253. "Angular Distribution of Pions Scattered by Hydrogen." With H. L. Anderson, D. E. Nagle, and G. B. Yodh. *Phys. Rev.* 86 (1952): 793–94 (letter). *FP*, 2:850–52.
254. "Scattering and Capture of Pions by Hydrogen." With H. L. Anderson. *Phys. Rev.* 86 (1952): 794 (letter). *FP*, 2:853–54.
255. "Report on Pion Scattering." Excerpts from the Proceedings of the Third Annual Rochester Conference, December 18–20, 1952. *FP*, 2:855–60.
256. "Numerical Solution of a Minimum Problem." With N. Metropolis. Los Alamos Document LA–1492, November 19, 1952. *FP*, 2:861–70.
257. "Angular Distribution of Pions Scattered by Hydrogen." With H. L. Anderson, R. Martin, and D. E. Nagle. *Phys. Rev.* 91 (1953): 155–68. *FP*, 2:871–901.
258. "Nucleon Polarization in Pion-Proton Scattering." *Phys. Rev.* 91 (1953): 947–48. *FP*, 2:902–5.
258a. "Scattering of 169 and 192 MeV Pions by Hydrogen." With R. L. Martin and D. E. Nagle. (Abstract). *Phys. Rev.* 91 (1953): 467.
259. "Scattering of Negative Pions by Hydrogen." With M. Glicksman, R. Martin, and D. E. Nagle. *Phys. Rev.* 92 (1953): 161–63. *FP*, 2:906–12.
260. "Phase Shift Analysis of the Scattering of Negative Pions by Hydrogen." With N. Metropolis and E. F. Alei. *Phys. Rev.* 95 (1954): 1581–85. *FP*, 2:913–22.
261. "Magnetic Fields in Spiral Arms." With S. Chandrasekhar. *Astrophysical Journal* 118 (1953): 113–15. *FP*, 2:923–30.
262. "Problems of Gravitational Stability in the Presence of a Magnetic Field." With S. Chandrasekhar. *Astrophysical Journal* 118 (1953): 116–41. *FP*, 2:931–59.
263. "Multiple Production of Pions in Pion-Nucleon Collisions." *Academia Brasileira de Ciências* 26 (1954): 61–63. *FP*, 2:960–63.
264. "Multiple Production of Pions in Nucleon-Nucleon Collisions at Cosmotron Energies." *Phys. Rev.* 92 (1953): 452–53. Errata Corrige, *Phys. Rev.* 93 (1954): 1434–35. *FP*, 2:964–69.
265. "Galactic Magnetic Fields and the Origin of Cosmic Radiation." *Astrophysical Journal* 119 (1954): 1–6. *FP*, 2: 970–77.
266. "Studies of Nonlinear Problems." With J. Pasta and S. Ulam. Los Alamos Document LA–1940, May 1955. *FP*, 2:978–88.

Bibliography

267. "Polarization of High Energy Protons Scattered by Nuclei." *Nuovo Cimento* 2 (1954): 407–11. *FP*, 2:989–93.
268. "Polarization in the Elastic Scattering of High Energy Protons by Nuclei." Private Communication, March 24, 1954. *FP*, 2:994–95.
269. "Physics at Columbia University—The Genesis of the Nuclear Energy Project." *Physics Today* 8 (November 1955): 12–16. *FP*, 2:996–1003.
270. "Lectures on Pions and Nucleons." Edited by B. T. Feld. *Nuovo Cimento* 2, supp. (1955): 17–95. *FP*, 2:1004–76.

Books

The Collected Works of Enrico Fermi. 2 vols. Chicago and Rome: University of Chicago Press, Accademia Nazionale dei Lincei, 1962, 1965.

Conferenze di Fisica Atomica (Fondazione Donegani). Rome: Accademia Nazionale dei Lincei, 1950.

Elementary Particles. New Haven: Yale University Press, 1951. Italian translation: *Particelle elementari*, translated by P. Caldirola. Milan: Einaudi, 1952.

Fisica ad uso dei Licei, 2 vols. Bologna: Zanichelli, 1929.

Fisica per le Scuole Medie Superiori. With Persico. Bologna: Zanichelli, 1938.

Introduzione alla fisica atomica. Bologna: Zanichelli, 1928.

Molecole e cristalli. Bologna: Zanichelli, 1934. German translation: *Moleküle und Kristalle*, translated by M. Schön and K. Birus. Leipzig: Barth, 1938. English translation: *Molecules, Crystals and Quantum Statistics*, translated by M. Ferro Luzzi. New York: Benjamin, 1966.

Notes on Quantum Mechanics. Chicago: University of Chicago Press, 1961.

Nuclear Physics. A course given at the University of Chicago. Notes compiled by J. Orear, A. H. Rosenfeld, and R. H. Schluter. Chicago: University of Chicago Press, 1949.

Thermodynamics. New York: Prentice-Hall, 1937. Italian translation: *Termodïnamica*, translated by A. Scotti. Turin: Boringhieri, 1958.

Notes on Thermodynamics and Statistics. Chicago: University of Chicago Press, 1966.

Index

Abelson, Philip H. (b. 1913), discovers neptunium, 117
Accademia d'Italia, 61
Adamson, Keith F., 114
Aeby, Jack, 148
Agnew, Harold M. (American physicist), studies with Fermi, 168
Alfvén, Hannes (Swedish physicist), discusses magnetohydrodynamics with Fermi, 175
Allison, Samuel K. (American physicist, 1900–65), 166, 167, 184; helps direct nuclear work in Chicago, 142, 143; participates in Trinity test, 146, 147; director of Institute for Nuclear Studies of the University of Chicago, 156–58
Amaldi, Ginestra, 81, 97
Amaldi, Edoardo (Italian physicist, b. 1908), vii, 215, 220; meets Fermi, 40; member of the Rome group, 49, 50, 55, 57, 58, 60, 68; participates in neutron work, 74, 77, 78, 79, 80, 81, 82; shares in neutron patents, 83; is left to work alone with Fermi, 86, 89; professor in Rome, 91, 97, 98, 105, 131, 175; welcomes Fermi back to Italy, 176
Amaldi, Ugo (Italian mathematician, 1875–1957, 40
Amidei, Adolfo, teaches and advises the boy Fermi, 8, 12
Anderson, Herbert L. (American physicist, b. 1914), 223, 226, 234, 237; studies with Fermi, 106, 108; collaborates on neutron work at Columbia and Chicago, 111, 112, 121, 126; helps start the Chicago pile, 127, 128; works at Los Alamos 131, 141; participates in Trinity test, 146; professor at Chicago, 167, 178
Anderson, Jean (wife of Herbert L. Anderson), 167
Appell, P., 16, 17, 191, 192
Argo, Harold V. (American physicist), studies with Fermi, 168
Ariosto, Lodovico (Italian poet, 1474–1533), 5
Armellini, Giuseppe (Italian astronomer, 1887–1958), 201
Astin, Allen V. (Director of the National Bureau of Standards, b. 1904), 180
Atomic Energy Commission (AEC), 84–85

Bacher, Robert F., (American physicist, b. 1905), works at Los Alamos, 135
Bainbridge, Kenneth T., (American physicist, b. 1904), works at Los Alamos, 144
Barbi, Michele (Italian philologist, 1867–1941), 14
Bard, Ralph A., 152
Barnard, Chester I., 159

Baudino, John (Fermi's bodyguard), 126, 139
Beams, Jesse W., (American physicist, b. 1898), 115
Becker, H., 68
Bergonzi, Giulia (Fermi's grandmother, b. 1830), 2
Bernardini, Gilberto (Italian physicist, b. 1906), 57, 176
Beta-ray theory, 71
Bethe, Hans A., (German-born American physicist, b. 1906), 117; visits Rome, 55, 59; works at Los Alamos, 135, 137; works on hydrogen bomb, 166, 174
Betti, Enrico (Italian mathematician, 1823–92), 14
Bhabha, Homi J. (Indian physicist, 1909–66), visits Rome, 59
Bianchi, Luigi (Italian mathematician, 1856–1928), 9, 14
Bjerge, T. (Danish physicist), works on neutrons, 77, 78, 82, 87
Blackett, Patrick M. (British physicist), 70
Blaserna, Pietro (Italian Senator and physicist, 1836–1918), 29
Bloch, Felix (Swiss-born American physicist, b. 1905), visits Rome, 59
Bohr, Aage (Danish physicist, son of Niels Bohr, b. 1922), works at Los Alamos, 139
Bohr, Niels (Danish physicist, 1885–1962), 17, 23, 32, 35, 170, 199, 205, 220, 221, 223, 225; attends Como Conference, 50; attends Solvay Conference, 70; Compound nucleus theory, 86, 89–90; informs Fermi of Nobel Prize, 97, 99; brings news of fission to U.S., 105-6; visits Los Alamos, 139; meets Churchill, 150
Boltzmann, Ludwig (Austrian physicist, 1844–1906), 17, 42, 43, 197
Bompiani, Enrico (Italian mathematician, b. 1889), 40
Bonomi, Ivanoe (Italian political leader, 1873–1951), 29
Booth, Eugene T. (American physicist, b. 1912), 108; separates uranium isotopes, 117; work at Columbia University, 225, 226, 230
Borden, William L. (former executive director of the staff of the Joint Committee on Atomic Energy), denounces Oppenheimer, 180
Born, Max (German physicist, 1882–1970), 58, 170, 183; receives Fermi as post-doctoral fellow, 32; attends Como Conference, 50
Börnstein, Landolt, physical tables used by Fermi, 56
Bose-Einstein: statistics, 42–43; condensation, 170
Bothe, Walther (German physicist, 1891–

269

Index

1957), 217; prepares discovery of neutron, 68–69; attends Solvay Conference, 70
Bowen, Harold G. (U.S. Admiral), allocates money to uranium project, 111
Bragg, Sir William L. (English physicist, b. 1890), 20
Breit, Gregory (American physicist, b. 1899), 220; work on neutron resonances, 89; member of uranium committee, 114–15; works on nuclear explosion, 133
Bretscher, Egon (Swiss-born English physicist, b. 1901), 141
Briggs, Lyman J. (American physicist, director National Bureau of Standards, member of uranium committee, 1874–1963), 114–15, 119, 123, 234
Broglie de, Louis V. (French physicist, b. 1892), 42, 44; participates in Solvay Conference, 70
Brueckner, Keith (American physicist, b. 1924), proposes nucleon-pion resonance, 179
Buckley, Oliver E., 163
Bush, Vannevar (American engineer and administrator, b. 1890), 123, 124; proposes National Defense Research Committee (NDRC), 114–15; creation of OSRD, 119; member of the Interim Committee, 151–52
Byrd, Richard E. (American admiral and polar explorer, 1888–1957), 104
Byrnes, James F., (American politician), 152, 154

Cantone, Michele (Italian physicist, 1857–1932), 45, 210, 212, 213
Caraffa, Andrea, his book studied by Fermi, 1, 2, 8
Carducci, Giosuè (Italian poet, 1835–1907), 14
Carrara, Nello (Italian physicist, b. 1899), 20
Carrelli, Antonio (Italian physicist), 210
Castelnuovo, Guido (Italian mathematician, 1865–1952), 40, 41; opens his home to Fermi, 14, 26; questions on quantum mechanics, 64; receives Fermi upon his return to Italy, 177
Cesalpino, Andrea (Italian anatomist and physiologist 1519–1603), 13
Cesàro, Ernesto (Italian mathematician, 1859–1906), 9
Chadwick, Sir James (British physicist, b. 1891): discovers neutron, 68–69; attends Solvay Conference, 70
Chamberlain, Owen (American physicist, b. 1920), 167
Chandrasekhar, Subrahmanyan (Indian-born American theoretical astrophysicist, b. 1910), reports Fermi's account of slow neutron discovery, 79–80
Chang, W. Y., 169
Chew, Geoffrey (American physicist, b. 1924), 137, 168
Churchill, Sir Winston (1874–1965), 122, 148, 150–52, 155
Chwolson, O. D., physics treatise, 189
Ciccone, Anna (Italian physicist), 21
Clayton, W. L. (US Assistant Secretary of State), participates in Interim Committee, 152
Cockcroft, Sir John D. (British Physicist and administrator, 1897–1967), participates in Solvay Conference, 70
Cohen, Karl, works on separation of uranium isotopes, 117

Colby, W. F. (American physicist), 63
Compton, Arthur H. (American physicist, 1892–1962), 142, 149, 150, 231, 236; views of U.S. scientists on nuclear energy, 110, 119; visits Fermi at Columbia, 120; transfers reactor project to Chicago, 121; works on uranium committee, 123–24; present at first criticality experiment, 127, 129; present at start of Hanford reactor, 143; member of panel of Interim Committee, 152; promotes post-war Chicago institutes, 156
Compton, Karl T. (American physicist, brother of A. H. Compton, 1887–1954), 115; member of Interim Committee, 152
Conant, James B. (American chemist and educator, b. 1893), 115, 144; participates in uranium committees, 123–24; informed of success of first reactor, 129; is appointed to Interim Committee, 152; is appointed to GAC, 163
Condon, E. U. (American physicist, b. 1902), 135
Conversi, Marcello (Italian physicist, b. 1917), work on mesons, 174
Corbino, Epicarmo (Italian economist, brother of Orso Mario Corbino, b. 1890), 27
Corbino, Leone (brother of Orso Mario Corbino), 27, 28
Corbino, Lupo (brother of Orso Mario Corbino), 27
Corbino, Orso Mario (Italian physicist, 1876–1937), 57, 207, 210, 212, 213; biography of, 26–30; recognizes and supports the young Fermi, 31–32, 34, 37, 40, 42; promotes the cause of theoretical physics in Italy, 44–46; wants to establish a physics center in Rome, 47, 49–50, 53; is invited to Solvay Conference, 54; secures Fermi's appointment to the Accademia d'Italia, 61–62; attitude toward quantum mechanics, 64–65; speaks on future development of physics, 65–68; prematurely announces transuranic elements, 76–77; suggests patenting slow neutrons, 83; death, 91; influence on Fermi, 93, 101
Corbino, Vincenzo (father of Orso Mario Corbino), 27
Creutz, Edward C. (American physicist), 228
Crommelin C. A. (Dutch physicist), 206
Curie, Marie (Polish-born French physicist and chemist, 1867–1934), vii, 74, 105; participates in Solvay Conference, 70

Dante Alighieri (1265–1321), Fermi's knowledge of, 5, 13
D'Agostino, Oscar (Italian chemist), participates in neutron work, 74, 83, 87, 215
D'Ancona, Alessandro (Italian literary historian, 1835–1914), 14
Dean, Gordon (AEC Commissioner), 165
Debye, Peter J. (Dutch-born physicist, 1884–1966), 54, 58
De Tivoli, Arnoldo, 212
Dini, Ulisse (Italian mathematician, 1845–1918), 10, 14
Dirac, Paul A. M. (English physicist, b. 1902), 75, 184; Fermi's study of, 23, 44, 52, 55, 64; discovers independently Fermi's statistics, 44; attends Solvay

Index

Conference, 70; creation and destruction operators, 71; monopole, 177
Dollfuss, Engelbert (Austrian Chancellor, 1892–1934), murder of, 94
Donegani lectures, given by Fermi, 176
Dreyfus, Alfred (French army officer, 1859–1935), 181
Drude, Paul (German physicist, 1863–1906), 48
Du Bridge, Lee A. (president Cal. Tech., member of GAC, b. 1901), 163; attends philosophical seminars in Aspen, 175
Dunning, John Ray (American physicist, b. 1907), 219, 220, 223, 225, 226, 230; works on neutrons at Columbia, 105–6, 108; separates uranium isotopes, 117

Ehrenfest, Paul (Austrian-born Dutch physicist, 1880–1933): befriends Fermi in Leyden, 34, 36; attends Ann Arbor summer school, 63
Einaudi, Renato (Italian mathematician), 58
Einstein, Albert (1879–1955), 149, 227; develops Bose-Einstein statistics, 42–43; alerts President Roosevelt to atomic energy, 112–14
Eisenhower, Dwight D. (1890–1969), 179; suspends Oppenheimer's clearance, 180, 182
Ellett, Alexander (American physicist, b. 1894), experiments on depolarization of resonance light, 38
Ellis, C. D. (British physicist), 68; participates in Solvay Conference, 70
Enriques, Federigo (Italian mathematician, 1871–1946), 24, 26, 41, 208
Enriques, Giovanni (Italian industrialist, son of Federigo, b. 1905), friend of Fermi, 24, 49–50
Eredia, Filippo (Italian meteorologist), Fermi's teacher in high school, 189
Evans Ward V., (American chemist), votes for Oppenheimer in security hearings, 181, 182

Fallopio, Gabriello (Italian anatomist, 1523–62), 13
Fano, Ugo (Italian-born American physicist, b. 1912), 58
Fantappiè, Luigi (Italian mathematician, 1901–56), 78
Farrell, Thomas (assistant to Leslie Groves), 125
Farwell, George (American physicist), 168
Fedele, Pietro (Italian minister of education), 209
Federzoni, Giovanni, Fermi's high school Italian teacher, 6
Feenberg, Eugene (American physicist), visits Rome, 59
Feld, Bernard T. (American physicist, b. 1919), member of Fermi's Columbia group, 121
Fermi, Alberto (E. Fermi's father, 1857–1924), 3–5, 8, 33
Fermi, Enrico (1901–54), vii, viii, 2; pre-school years, 5; elementary and high school, 6–11; meets Persico, 7; attends Scuola Normale Superiore at Pisa, 12–15; meets Rasetti, 15; reviews his knowledge of physics, 17–18; Doctoral dissertation, 19–21, 22; first publication, 21; scientific style, 22, 55, 59, 104; work on relativity, 24; meets Corbino, 26, 30, 31; fellowship at Göttingen, 32; death of parents, 33; temporary job in Rome, 34; collision theory, 34–36; publication policy, 35; meets Ehrenfest, 36; temporary job in Florence, 37–40; competes for a chair of mathematical physics, 40–42; discovers Fermi statistics, 42–44; wins competition for a chair of theoretical physics, 45; actions to foster study of modern physics, 46–48; participates in Como physics conference, 50; tutors selected students, 50–52; Thomas-Fermi atom, 53; lectures in Leipzig, 54; work on electrodynamics, 55; attends Solvay conferences, 54, 70; Rome school, 49–53, 55, 59; appointed to Accademia d'Italia, 61; visits the U.S., 63; studies quantum mechanics, 64; argues need of new departures in physics, 65, 67; starts nuclear studies, 68; hyperfine structure studies, 68; reports on nucleus at Paris, 69; names neutrino, 70; theory of beta decay, 70-72; finds artificial radioactivity by neutron bombardment, 73–77; trip to South America, 77; discovery of slow neutrons, 79–82, 87, 90; patents, 83–85, 244n; political attitudes, 93, 101–4; decides to emigrate, 97; receives Nobel Prize, 97–99; arrives in New York, 100; professor at Columbia, 104; efforts at Americanization, 104; learns of fission, 105; starts working on chain reaction, 106–9; alerts U.S. government, 111; attends meetings of uranium committee, 114; establishes fundamentals of chain reaction, 116; discusses plutonium, 118; transfers Columbia work to Chicago, 120–23; relations with Groves, 125; works on pile at Chicago, 126–29; applies chain reaction to physics, 130–32; helps with engineering problems 132; visits and consults at Los Alamos, 135–40; transfers to Los Alamos, 141; helps start Hanford production reactor, 142–44; participates in Trinity test of nuclear explosion, 146–48; participates in formulation of "prospectus on nucleonics," 150–51; consultant of Interim Committee, 152–54; decides to leave Los Alamos, 155; accepts professorship at Chicago, relation to Allison, 156–58; political suggestions, 159, 161–63; member of the GAC, 163–66; decides to study particle physics, 166; establishes a school at Chicago, 168–71; attends conferences on electrodynamics, 172–74; theoretical work on mesons, cosmic rays, 174, 175, 177; visits Italy, 176, 183; experiments on pions, 178, 182; president of the American Physical Society, 179; acts in Astin incident, 180; testifies in Oppenheimer affair, 180–82; death, 183, 184, 185; correspondence with Persico, 189–213; Speeches, 222, 231
Fermi, Giulio (E. Fermi's brother, 1900–15), 5, 6, 7, 11, 33
Fermi, Giulio (son of Enrico and Laura Fermi, b. 1936), 61, 98, 167
Fermi, Ida de Gattis (mother of E. Fermi, 1871–1924), 4, 7, 3
Fermi, Laura (wife of E. Fermi), vii, 11, 15, 30, 33, 61, 77, 98, 104, 130, 139
Fermi, Maria(E. Fermi's sister, 1899–1964), 5, 32, 33, 34, 204
Fermi, Nella (daughter of Enrico and Laura Fermi, b. 1931), 61, 98, 167

271

Index

Fermi, Stefano (grandfather of E. Fermi, 1818–1905), 2, 3
Feynman, Robert P. (American physicist, b. 1918): works at Los Alamos, 137; participates in conferences on electrodynamics, 173, 174
Fink, G. (American physicist), 219
Flanders, Donald A. (American mathematician), 137
Fleischmann, Rudolph, 220
Franck, James (German-born American physicist, 1882–1964): professor in Göttingen, 32; works at Chicago Metallurgical Laboratory, 132; concern about use of atomic energy, 150, 151; writes Franck report, 153
Frisch, Otto R. (Austrian-born English physicist), 106, 223; neutron work, 90; fission work, 99
Fubini, Eugenio (Italian-born American engineer), studies in Rome, 58
Fubini, Guido (Italian mathematician, 1879–1943), 14

Galeotti, Gino (Italian pathologist, 1867–1921), uncle of Franco Rasetti, 15
Galileo Galilei (1564–1642), 5, 13, 26, 37
Gamow, George (Russian-born American physicist, 1904–68), participates in Solvay Conference, 70
Garbasso, Antonio (Italian physicist, director of physics laboratory in Florence, 1871–1933), 210, 212, 213; instrumental in hiring Fermi and Rasetti, 37, 45; polemic with Corbino, 67–68
Garwin, Richard L. (American physicist, b. 1928), studies with Fermi, 168
Gemelli, Father Agostino (president of Catholic University of Milan), 209
General Advisory Committee (GAC), 162
General Electric Company, early attitude to nuclear research, 84
Gentile, Giovanni (Italian philosopher, 1875–1944), 206, 208; has Fermi as collaborator in Enciclopedia Italiana, 103
Gentile, Giovanni, Jr. (Italian physicist, 1906–42), works at Rome, 58
Giannini, G. M. (Italian-American businessman), 85; studies at Rome, 51; shares in neutron patents, 83; welcomes Fermi in New York, 100
Giolitti, Giovanni (Italian political leader, 1842–1928), appoints Corbino senator, 29
Giorgi, Giovanni (Italian electrical engineer, 1871–1950), wins competition for chair of mathematical physics, 41, 42
Giusti, Giuseppe (Italian poet, 1809–50), 14
Glasoe, Norris G. (American physicist), works with Fermi at Columbia, 108
Goldberger, Marvin (American physicist), studies with Fermi, 168
Goudsmit, Sam (Dutch-born American physicist, b. 1902), visits Rome, 59; friendship with Fermi at Leyden and Ann Arbor, 63
Gray, Gordon, member of Personnel Security Board of AEC, 181, 182
Greenewalt, Crawford H. (chemical engineer, later President of Du Pont, b. 1902), 132; present at criticality experiment, 129; present at start of Hanford reactor, 144
Groves, Leslie R. (American general, commander of Manhattan District, b. 1896), 124–26, 132, 133, 143; selects Los Alamos site, 134; present at Trinity test, 146, 147; attitude to nuclear problems, 150, 152, 158
Guglielmo, judge in Cagliari competition, 41
Gunn, Ross (physicist at Naval Research Laboratory during World War II): appreciates military importance of nuclear work, 111; member of uranium committee, 115
Gustav V, king of Sweden, presents Nobel Prize to Fermi, 98, 99

Hahn, Otto (German physical chemist, 1879–1968), 217, 223; cordial relations with Rome, 92; discovers fission, 99; discovers uranium, 117, 239
Halban, Hans von (Austrian physicist), 224; reports about French neutron work, 90; studies chain reaction, 107
Hanle, Wilhelm (German physicist, b. 1901), work on optical resonance, 38
Harrison, George L., on use of atomic weapon against Japan, 153
Heisenberg, Werner (German physicist, b. 1901), 23, 47, 58, 60, 88, 209; meets Fermi in Göttingen, 32, 33; discovers quantum mechanics, 44; attends Como Conference, 50; quantum electrodynamics work, 55; Solvay Conference, 70
Hinton, Joan, assistant to Fermi in Los Alamos, 141
Hitler, Adolf (1889–1945): visit to Rome, 94–95; invasions, 114; hate of Hitler spurs atomic work, 124, 145
Hooper, Admiral S.C., alerted to nuclear military applications, 110, 111
Hoover, Commander Gilbert C., on uranium committee, 114
Hoover, J. Edgar (b. 1895), receives report on Oppenheimer, 180
Houzeau, J. C., 201
Hughes, D. J. (American physicist), signs Franck report, 153
Hutchins, Robert M. (president of University of Chicago, b. 1899): supports new institutes at University of Chicago, 156; calls conference on atomic energy, 159
Huxley, Aldous L. (English novelist and critic, 1894–1963), Fermi reads writings of, 90, 103

Imprescia, Rosaria (b. 1858), mother of Orso Mario Corbino, 27
Inoue, T. (Japanese physicist), hypotheses on mesons, 174
Interim Committee, 152
Isotope separation, 117
Ivanenko, D. (Russian physicist), proposes nuclear model, 70

Jeffries, Zay (metallurgical engineer), committee issues a prospectus on nucleonics, 150, 151
Jensen, Hans D. (German physicist, b. 1907), discovers nuclear shells, 172
Jewett, Frank B. (president National Academy of Sciences), 115
Johnson, Edwin C. (senator of Colorado), proposes atomic energy bill, 161
Joliot, Frédéric (French physicist, 1900–58), 214, 224; works on neutrons, 68–

272

Index

69; attends Solvay Conference, 70; discovers artificial radioactivity, 72–73; studies chain reaction, 107, 113
Joliot, Irène Curie (French chemist, 1897–1956), 214; works on neutrons, 68–69; attends Solvay Conference, 70; discovers artificial radioactivity, 72–73
Jordan, Pascual (German physicist), meets Fermi in Göttingen, 32, 33; quantum electrodynamics work, 55; destruction and creation operators, 71

Kapitza, Piotr L. (Russian physicist), 90
Kennedy, Joseph W. (American chemist, 1916–57), work on plutonium, 118, 119
King, Percival (American physicist, b. 1906), works with Fermi at Los Alamos, 141
Kistiakowski, George B. (Russian-born American physicist, b. 1900), works on explosives at Los Alamos, 144
Klein, Oscar (Swedish physicist), creation and destruction operators, 71
Kowarski, Lev N. (Russian-born French physicist, b. 1907), studies chain reaction, 107
Krewer, S. E., helps Szilard in experiment, 108
Kronig, R. de L. (American physicist, b. 1904), hikes in the Alps with Fermi, 40
Kusch, Polykarp (American physicist, b. 1911), measures magnetic moment of electron, 173

Lamb, Willis E. (American physicist, b. 1913), 223; hears of discovery of fission, 105; Lamb-shift, 173–74
Langmuir, Irving (American chemist, 1881–1957), 231
Laska, collection of formulas by, 56
Lawrence, Ernest O. (American physicist, 1901–58), 150; attends Solvay Conference, 70; meets Fermi in Berkeley, 116; interest in atomic project, 117–19; electromagnetic separation of isotopes, 123; works on uranium committees, 124; technical panel of interim committee, 152; supports May-Johnson bill, 161–62; advocates crash program on hydrogen bomb, 165
Lazarus, David (American physicist, b. 1921), studies with Fermi at Chicago, 168
Lea, D. E. (British physicist), 219; finds gamma rays on neutron capture, 82
Lee, Tsung Dao (Chinese-born American physicist, b. 1926), studies with Fermi at Chicago, 168
Leverett, Miles C., 244n
Levi-Civita, Tullio (Italian mathematician, 1873–1941), 23, 199, 207, 210, 213; befriends Fermi, 26, 36, 40; referee for Fermi, 41
Lilienthal, David E. (AEC commissioner, b. 1899), 159, 163
London, Fritz (German-born American physicist, 1900–54), visits Rome, 59
Lorentz, Hendrik Antoon (Dutch physicist, 1853–1928), 28, 177; theory of electrons, 17; attends Como Conference, 50
Lo Surdo, Antonino (Italian physicist, 1880–1949), 49, 53, 212; hostile toward Fermi, 48; hampers Fermi nomination to the Accademia dei Lincei, 61; director of the Institute in Rome, 91

Lundby, Arne, experiments on pions with Fermi, 178

Macaluso, Damiano (Italian physicist, 1845–1932), Corbino's teacher, 28
MacMahon bill, 162
Maggi, Gian Antonio (Italian mathematician, 1856–1937), 45, 210
Majorana, Ettore (Italian physicist, 1906–38): student in Rome, 51, 54, 55, 59; ideas on the neutron, 68–70
Majorana, Quirino (Italian physicist, 1871–1957), 45, 210, 212, 213
Malpighi, Marcello (Italian anatomist, 1628–94), 13
Manley, John (American physicist), secretary of GAC, 135, 164
Manzoni, Alessandro (Italian poet, 1785–1873), 3, 166
Marcolongo, Roberto (Italian mathematician, 1861–1943), Fermi's referee, 41
Marshak, Robert (American physicist), attends Shelter Island Conference, 174; organizes High Energy Conferences, 177
Marshall, General George C. (1880–1959), 124; meets with Interim Committee, 152, 153
Marshall, John (American physicist, b. 1925), 171; professor at Chicago, 167; directs construction of Chicago cyclotron, 178
Marshall, Leona Woods (American physicist, b. 1919), 171; works on neutrons with Fermi, 131; joins Chicago faculty, 167
Martin, Ronald L. (American physicist, b. 1922), works on neutrons with Fermi, 178
Matteotti, Giacomo (Italian politician, 1885–1924), victim of fascism, 93
Maxwell, James Clark (Scottish physicist, 1831–79), 185, 194
Mayer, Maria Göppert (German-born American physicist, b. 1906), discovers nuclear shells, 172
May-Johnson bill, 161, 162
McCarthy, Senator Joseph (1908–57), 179
McMillan, Edwin M. (American physicist, b. 1907), discovers neptunium, 117
MED, Manhattan District of the Corps of Engineers, 124
Meitner, Lise (German physicist, 1878–1968), 217, 223; receives Rasetti in her laboratory, 58, 68; attends Solvay Conference, 70; victim of Nazism, 92; work on fission, 99; discovers U 239, 117
Metropolis, Nicholas (American physicist, b. 1915), work on computers, 179
Millikan, Robert A. (American physicist, 1868–1953), receives Rasetti in his laboratory, 58
Mitchell, Dana P. (American physicist), 105, 219
Møller, Christian (Danish physicist, b. 1904), 59
Moon, P. B. (English physicist, b. 1907), 83, 87, 218, 220
Moore, Henry (British sculptor, b. 1898), 129
Moore, Thomas V. (chairman of engineering council, Metallurgical Laboratory), 130
Morgan Thomas, member of Personnel Security Board of AEC, 181

273

Index

Morrish, Allen H. (Canadian physicist), studies with Fermi, 168
Mortara, Nella, 212
Mossotti, Ottaviano (Italian physicist, 1791–1863), 14
Murphree, Eger V., petroleum executive on uranium committee, 123, 124
Mussolini, Benito (1883–1945), 166; appoints Corbino to his cabinet, 29; march on Rome, 30–31; appoints Fermi to Accademia d'Italia, 61–62; disastrous policies, 93–96

Nagle, Darragh E. (American physicist, b. 1919), studies and works with Fermi, 168, 178
NDRC (National Defense Research Committee), 115, 120
Neddermeyer, Seth (American physicist, b. 1907), proposes implosion method, 137, 144
Nernst Walter H. (German chemist, 1864–1941) 193
Neumann von, John (Hungarian-born American mathematician, 1903–57), 107; consultant at Los Alamos, 137, 138, 140, 141; works on hydrogen bomb, 166
Nichols, Colonel Kenneth D. (aide to Groves), 125
Nickson, J. J., signs Franck report, 153
Nier, Alfred O. (American physicist), 225
Noddack, Ida (German chemist), predicts uranium fission, 76
Nordheim, Lothar W. (German-born American physicist, b. 1899), visits Rome, 59

Occhialini, Giuseppe (Italian physicist, b. 1907), 37, 78, 176
Oppenheimer, J. Robert (American physicist, 1904–68), 150, 231; directs bomb study in Berkeley, 123; director at Los Alamos, 133–44; participates in Trinity test, 146, 147, 148; member of technical panel of Interim Committee, 152; endorses May-Johnson bill, 161; chairman of GAC, 162–64; director of Institute for Advanced Study, 173; security hearings, 180–82, 184
Orear, Jay (American physicist), studies with Fermi in Chicago, 168, 169
Ornstein L. S. (Dutch physicist), 209
OSRD (Office of Scientific Research and Development), 119, 123
Ostwald, Wilhelm (German chemist, 1853–1932), 193

Pancini, Ettore (Italian physicist), work on mesons, 174
Parsons, Admiral William S., work at Los Alamos, 144
Patterson, Robert P. (Secretary of War), 161
Pauli, Wolfgang (Austrian-born Swiss physicist, 1900–58), 184; discovers exclusion principle, 42–43; participates in Como Conference, 50; work on quantum electrodynamics, 55, 60; participates in Solvay Conference. 70; formulates neutrino hypothesis, 71
Peano, Giuseppe (Italian mathematician, 1888–1932), 10
Pegram, George B. (American physicist, 1876–1958), 108, 118, 219, 220, 222, 229, 234; welcomes Fermi to New York, 100; head of physics department at Columbia, 104–5; alerts government on uranium problem, 110–11; participates in uranium committees, 114–15
Peierls, Rudolf (German-born British physicist), visits Rome, 59
Perrin, Francis (French physicist, b. 1901), participates in Solvay Conference, 70
Persico, Enrico (Italian physicist, 1900–69), vii, 189; boyhood friendship with Fermi, 7, 11; letters to Fermi, 12, 16, 18, 19, 21, 22, 34, 39, 189–213; propounder of theoretical physics in Italy, 40–41, 47, 48, 66; professor of theoretical physics in Florence, 44–45, 57; professor in Turin, 58; present at discovery of slow neutrons, 80
Philips Company, Eindhoven Netherlands, early appreciation of nuclear physics, 83, 84
Piccioni, Oreste (Italian-born American physicist, b. 1915), work on mesons, 174
Pierucci, Mariano (Italian physicist), 21
Pincherle, Leo (Italian-born British physicist), works at Rome, 58
Pittarelli, Giulio (Italian mathematician, 1852–1934), examines the boy Fermi, 13
Placzek, George (Czech-born American physicist, 1905–55): visits Rome, 59–60; reports on Copenhagen neutron work, 90; work on chain reaction, 108
Planck, Max (German physicist, 1885–1947): *Thermodynamics*, 16, 52, 193; reaction to wave mechanics, 44–45; attends Como Conference, 50
Pleijel, H. (Swedish physicist), 98
Poincaré, Henri (French mathematician, 1854–1912), *Théorie des tourbillons*, 16, 191
Poisson, S. D. (French mathematician, 1781–1840), 10, 12, 16; *Traité de mécanique*, 192
Polvani, Giovanni (Italian physicist), 21, 22, 200, 210
Pontecorvo, Bruno M. (Italian-born Russian physicist, b. 1913), 215; studies at Rome, 58, 78; works on neutrons, 79, 81–82; neutron patents, 83; goes to Russia, 85; goes to France, 87; welcomes Fermi in Italy, 176
Pontremoli, Aldo (Italian physicist), 45, 210
Preiswerk, Peter (b. 1907), reports neutron work in Paris, 90
Pucciantí, Luigi (Italian physicist, 1875–1946), 197; professor of physics at Pisa, 18, 20, 21, 26, 36, 37; spectroscopic work, 65

Rabi, I. I. (Austrian-born American physicist, b. 1898): molecular beams at Columbia, 105; consultant at Los Alamos, 135; member of GAC, 163, 165; attends electrodynamics conference, 173
Rabinowitch, Eugene (American biophysicist), signs Franck report, 153
Racah, Giulio (Italian-born Israeli physicist, 1909–65), visits Rome, 55, 57
Rajna, Pio (Italian philologist, 1847–1930), 14
Raman-effect, 49, 58, 60, 66
Rasetti, Adele (mother of Franco Rasetti), 15, 203
Rasetti, Franco (Italian physicist and paleontologist, b. 1901), 207, 208, 213 220; student at Pisa, 15, 16, 19, 20; works at Florence, 37, 38; propounder

274

Index

of modern physics in Italy, 47; experimental work in Rome, 49, 50, 52, 53, 54, 57, 60, 72; attends Fermi's seminars in Rome, 55; goes to Pasadena, 58, 63; advocates change of work in Rome, 65, 68; participates in neutron work, 72, 73, 74, 81, 82; shares in neutron patents, 83–85; goes to Columbia University, 87, 89, 103, 105; builds small accelerator in Rome, 91; farewell to Fermi upon leaving Italy, 97, 98

Reitz John R. (American physicist), studies with Fermi, 168

Respighi, Ottorino (Italian composer, 1879–1936), 78

Ricci-Curbastro, Gregorio (Italian mathematician, 1853–1925), 14, 23

Richardson, O. W., (English physicist, 1879–1959), electron theory of matter, studied by Fermi, 17

Riemann, Bernhard (German mathematician, 1826–66), 14, 170

Righi, Augusto (Italian physicist, 1850–1920), 28, 30

Ronchi, Vasco (Italian optician), 37

Roosevelt, Franklin D. (1882–1945), 110; alerted to nuclear problems, 112–14; establishes OSRD, 119; removes war restrictions on Italians, 122; receives reports on uranium work, 123; consults with Churchill, 151

Rosenbluth, Marshall N. (American physicist), studies with Fermi at Chicago, 168

Rosenfeld, Arthur H. (American physicist), studies with Fermi at Chicago, 168, 169

Rossi, Bruno (Italian-born American physicist, b. 1905), 57; present at discovery of slow neutrons, 80; works at Los Alamos, 140, 144; visit to Italy in 1949, 176

Rowe, Hartley, member of GAC, 163

Rutherford, Ernest (British physicist, 1871–1937), 205, 214; Fermi studies his books, 17, 68; attends Como Conference, 50, 65; attends Solvay Conference, 70; congratulates Fermi, 74, 75; presents neutron work to Royal Society, 77; recommends work as cure against despondency, 90, 92

Sachs, Alexander (American economist), 112, 114

Sackur, Otto (German physical chemist), calculates entropy constant, 36, 42

Sakata (Japanese physicist), hypothesis on mesons, 174

Sbrana, Francesco (mathematical physicist), 210

Schluter, R. A., studies with Fermi at Chicago, 168, 169

Schrödinger, Erwin (Austrian physicist, 1887–1916): Fermi studies his papers 23, 44, 45, 47, 51, 52, 64; attends Solvay Conference, 69; flight to Rome, 95

Schuyler, Captain G. L., 111

Schwinger, Julian (American physicist, b. 1918), participates in conference on electrodynamics, 174

Seaborg, Glenn T. (American chemist, b. 1912): work on plutonium, 118, 119; signs Franck report, 153; member of GAC, 163, 165

Segrè, Emilio G. (Italian-born American physicist, b. 1905), 215, 219, 220; meets Fermi, 2, 50; meets Rasetti, 49; attends Como Conference, 50; introduces Majorana to Fermi, 51; work on neutrons, 73, 74, 75, 77, 80, 82; studies with Fermi, 50, 55, 68; works with P. Zeeman and O. Stern, 58; goes to Ann Arbor, 63; visits Cambridge, England, 77, 78; shares in neutron patent, 83; professor in Palermo, 87; emigrates to U.S., 98; work with Fermi in Berkeley, 116; work on plutonium, 118; works in Los Alamos, 135, 137, 139, 140, 143, 144; participates in Trinity test, 146, 147; return to Italy, 176; last scientific discussion with Fermi, 183; last visits to Fermi, 184

Sella, Alfonso (Italian physicist), 29

Selove, Walter (American physicist), studied with Fermi at Chicago, 168

Serber, Robert (American physicist, b. 1909), prepares Los Alamos primer, 135; 137

Severi, Francesco (Italian mathematician), 14

Shotwell, J. T., 231

Slack, Francis G., works on neutrons at Columbia, 108

Slow neutrons, discovery of, 80

Smith, Cyril S. (British-born American metallurgist, b. 1903): professor at Chicago, 156; member of GAC, 163, 167

Smyth, Henry De Wolf (American physicist, b. 1898), 120, 160, 161, 231

Solvay Conferences, 54, 70

Somigliana, Carlo (Italian mathematician, 1860–1955), 41, 210

Sommerfeld, Arnold (German physicist, 1868–1952): Fermi studies his book, 18, 44, 47; attends Como Conference, 50; his school in Germany, 32, 59, 60

Statistics, discovery of, 42

Staub Hans (Swiss physicist), 144

Stearns, Joyce C. (American physicist), signs Franck report, 153

Steinberger, Jack (American physicist), studies at Chicago, 168

Stern, Otto (German-born American physicist, 1888–1969): work on entropy constant, 36, 42; molecular beams in Hamburg, 58

Sternheimer, Rudolph M. (American physicist), studies with Fermi in Chicago, 168

Stewart, Irvin, 123

Stimson, Henry L. (Secretary of War, 1867–1950): briefs President Truman on atomic bomb, 152; chairman of Interim Committee, 152, 153–55

Stone, R. S., 231

Stracciati, Corbino's high school teacher, 28

Strassmann, Fritz (German chemist), 217, 223; discovers fission, 99

Strauss, Lewis L. (American investment banker, chairman of AEC, b. 1896), 84, 165, 180; reaction of American business to atomic energy, 84; contact with Szilard, 107, 110; proposes award of Fermi prize, 147n

Styer, General, works on uranium committee, 124

Sugarman, Nathan (American chemist), on Chicago faculty, 167

Szilard, Leo (Hungarian-born American physicist, 1898–1964), 223, 224, 226, 230, 234; advocate of nuclear energy, 106, 107, 110, 115, 120; works at Columbia, 108, 111, 112; enlists Einstein's support, 112, 113; on uranium committee, 114; discontent with MED, 132;

275

Index

views on use of atomic energy, 149, 150, 151, 154; signs Franck report, 153

Taft, Horace D. (American physicist), studies with Fermi, 168
Taylor, Sir Geoffrey (British applied mathematician, b. 1886), consultant at Los Alamos, 138
Teller, Edward (Hungarian-born American physicist, b. 1908), 107; visits Rome, 59; member of uranium committee, 114; works at Los Alamos, 137, 139, 141; advocates hydrogen bomb, 165, 166; professor in Chicago, 167, 169; work on meson, 174
Tetrode (German physicist), work on entropy constant, 36, 42
Thomas-Fermi model, 53, 170
Thomson, J. J. (British physicist, 1856–1940), book on gas discharges, 48
Tieri, Laureto (Italian physicist), 200
Tillman, J. R. (British physicist), 83, 87, 218, 220
Tolman, Richard C. (American physicist, 1881–1948), 150
Townsend (British physicist), Fermi studies his book, 17
Trabacchi, Giulio Cesare (Italian physicist), 74, 205, 212, 221; helps the Rome group, 53; supplies sources for neutron work, 73; shares in neutron patent, 83
Trevisan, Cornelia, 208
Tricomi, Francesco (Italian mathematician), 40, 206
Truman, Harry, S. (b. 1884): briefed on atomic bomb, 152, 153; decision on using the bomb, 155; atomic legislation, 161, 162; decides to build hydrogen bomb, 165
Turkevich, Anthony (American chemist), professor at Chicago, 167
Tuve, Merle A. (American physicist), in uranium committee, 114, 115

Uhlenbeck, George (Dutch-born American physicist, b. 1900), friendship with Fermi, 34, 63
Ulam, Stanislaw M. (Polish-born American mathematician), work on hydrogen bomb, 166
Uranium committees, 123
Urey, Harold (American chemist, b. 1893), 226, 230, 231; professor at Columbia, 104, 105; on uranium committee, 114, 115, 123, 124; work on isotope separation, 117; preoccupation with future of atomic energy, 151; professor in Chicago, 156, 167

Villari, Pasquale (Italian historian, 1826–1917), 14
Viner, Jacob, 231
Vitelli, Girolamo (Italian philologist, 1849–1935), 14
Volta, Count Alessandro (Italian physicist, 1745–1827), 65, 116
Volterra, Vito (Italian mathematician, 1860–1940), 14, 210; recommends Fermi for a fellowship, 34; supports Fermi in a competition, 41

Wahl, Arthur C. (American chemist), work on plutonium, 119
Wallace, Henry A., participates in conference on nucleonics, 159
Warshaw, Stephen I. (American physicist), studies with Fermi, 168
Wataghin, Gleb (Russian-born Italian physicist), 77, 78, 176

Watson, General E. M., assistant to President Roosevelt, 114
Wattenberg, Albert P. (American physicist), studies with Fermi, 168
Weil, George L. (American physicist), member of Columbia neutron group, 121
Weisskopf, Victor F. (Austrian-born American physicist), works at Los Alamos, 137; works on mesons, 174
Weizsäcker, C. F. (German physicist), 34, 35, 114
Wentzel, Gregor (German-born American physicist, b. 1898), professor at Chicago, 168
Wertenstein (Polish physicist), questions on neutrino, 69
Westcott, C. H., work on neutrons, 77, 78, 82, 87
Weyl, Hermann (Swiss mathematician, 1885–1955), Fermi studies his books, 56
Wheeler, Archibald J. (American physicist, b. 1911), 143, 225, 231
Wick, Gian Carlo (Italian-born American physicist, b. 1909), 58, 59
Wigner, Eugene P. (Hungarian-born American physicist, b. 1902), 71, 107, 120, 169, 220, 227, 231; comments on Fermi's work, 55, 72; work on neutron resonances, 89; enlists Einstein's support for atomic work, 112; member of uranium committee, 114–15; participates in pile experiments at Chicago, 129; work with Du Pont, 132
Williams, E. J. (British physicist), collision method, 34, 35
Williams, John H. (Canadian-born American physicist): works at Los Alamos, 135; participates in Trinity test, 145
Willits, J. H., 231
Wilson, Owen M., 175
Wilson, Robert R. (American physicist, b. 1914), 135, 228
Wilson, Volney C. (American physicist), 128, 237
Wolfenstein, Lincoln (American physicist), studies with Fermi at Chicago, 168
Wood, Robert W. (American physicist, 1868–1955), experiments on depolarization of resonance light, 38
Woods Marshall, Leona. See Marshall, Leona Woods
Worthington, Hood, member of GAC, 163
Wu, Chien Shiung (Chinese-born American physicist, b. 1913), 143

Yang, Chen Ning (Chinese-born American physicist, b. 1922), describes Fermi's teaching at Chicago, 168, 169, 170
Yodh, Gaurang B. (Indian-born American physicist), studies with Fermi at Chicago, 168, 178
Yukawa, Hideki (Japanese physicist, b. 1907), 71, 178

Zanichelli (publishing house, Bologna, Italy), 47
Zeeman, Pieter (Dutch physicist, 1865–1943), 38, 58, 170, 207
Zinn, Walter H. (Canadian-born American physicist, b. 1906), 223, 226; participates in the work at Columbia, 111, 121; participates in the pile experiment at Chicago, 127–28; uses pile as research instrument, 131

Segré, Emilio.

Enrico Fermi, physicist.

70-3691

530.9
Se38e

LAWRENCE PUBLIC LIBRARY
LAWRENCE, KANSAS